SPRING DESIGN
AND
MANUFACTURE

FOR HOME MACHINISTS

Workshop Practice Series

TUBAL CAIN

© 1988, 2000, 2007, 2013, 2016, 2023, 2025 by Tubal Cain
and Fox Chapel Publishing Company, Inc.

All rights reserved. *Spring Design and Manufacture for Home Machinists* is a revised edition of *Spring Design & Manufacture*, published in the UK in 2007 by Special Interest Model Books.

ISBN 978-1-4971-0523-2
The Cataloging-in-Publication Data is on file with the Library of Congress.

To learn more about the other great books from Fox Chapel Publishing, or to find a retailer near you, call toll-free at 800-457-9112 or visit us at *www.FoxChapelPublishing.com*. You can also send mail to:
Fox Chapel Publishing
903 Square Street
Mount Joy, PA 17552.

We are always looking for talented authors.
To submit an idea, please send a brief inquiry to acquisitions@foxchapelpublishing.com.

Printed in China

© Special Interest Model Books
An imprint of Fox Chapel Publishers International Ltd.
20-22 Wenlock Road
London
N1 7GU

www.foxchapelpublishing.co.uk

First published 2025
Text copyright 2025 Tubal Cain
Layout copyright 2025 Special Interest Model Books

ISBN 978-0-85242-925-9

Tubal Cain has asserted his right under the Copyright, Design and Patents Act 1988 to be identified as the author.

All rights reserved. No part of this publication may be reproduced in any form, by print, photography, photocopying, microfilm, electronic file, online or other means without written permission from the publisher.

Printed and bound in China

Because working with springs and other materials inherently includes the risk of injury and damage, this book cannot guarantee that creating the projects in this book is safe for everyone. For this reason, this book is sold without warranties or guarantees of any kind, expressed or implied, and the publisher and the author disclaim any liability for any injuries, losses, or damages caused in any way by the content of this book or the reader's use of the tools needed to complete the projects presented here. The publisher and the author urge all readers to thoroughly review each project and to understand the use of all tools before beginning any project.

Contents

Chapter 1	Introduction	4
Chapter 2	Tension and Compression Spring Principles	6
Chapter 3	Compression and Tension Spring Design	18
Chapter 4	Worked Examples	30
Chapter 5	Winding Coil Springs	35
Chapter 6	Leaf Spring Principles	47
Chapter 7	Leaf Spring Design	54
Chapter 8	Making Leaf Springs	69
Chapter 9	Torsion Springs	74
Chapter 10	I.C. Engine Valve Springs	87
	Index	95

ACKNOWLEDGEMENT
I take this opportunity of publicly acknowledging the help of the late Mr. W. R. Berry, M.Sc., who first introduced me to the use of Nomograms in solving spring problems, almost exactly 50 years ago. Those who would like to construct their own charts can find the derivation in the appendix to his paper "Practical Problems in Spring Design", *Proc. I.Mech.E.,* Vol 139, 1938, p. 474.   –T.C.

CHAPTER 1

Introduction

When this work was first considered I must confess that I had a few doubts. First, because the design of small springs is always subject to uncertainty. The specification of spring wire bought in "retail" quantities is, inevitably, both variable and unknown. The tolerance on the wire diameter may vary from coil to coil. And, of course, it is not easy to wind a spring to a specified mean diameter, still less to wind a number of springs to identical dimension; few of us can afford proper spring-winding machines. On top of this, the theory itself can be imperfect. Though it has been refined over the years, and correction factors added as our experience has increased, the range of sizes, shapes, and materials is so great that theory is bound to be stretched a little at the extremes – the very large and the very small, the latter category including most of those which are used in model-making. This means that a trial spring is almost always necessary.

However, the use of preliminary calculation does at least give us some idea of where to start when making this trial spring, and in many cases, a single trial will suffice. This saves a great deal of uncertainty. And, moreover, it usually means that the final spring is likely to be "just what is needed" rather than "near enough". However, this did not remove my second, and more serious, cause for apprehension – that the inevitable "formulas" might frighten readers who are not familiar with them to not even try to use them. I urge you to set all fears behind you!

First, these formulas are *no more than models.* They express, in symbols, what is happening inside the spring, that is all. Consider: it is no more than common sense that the load which a coil spring can carry will depend on (a) the wire diameter (b) the diameter of the coils and (c) the strength of the material used. An equation such as that at (2) on page 6 merely sets these three factors in their correct relationship, and "puts the numbers in" so that the safe load can be worked out.

Second, I have reduced almost all the necessary formulas to charts, some of which I have used for over half a century, from the days when designing springs was part of my daily work. All you need is a ruler to use them! But I do urge you to check the result using one or more of the formulas, for whereas my own charts are a couple of feet tall, they have had to be reduced in size to get them into the page. In addition, I have tried in most cases to explain *why* the formulas are in the shape they are, both to help you understand them, and because this also helps us to decide what to do if the spring just won't fit into the space available. One of the crosses we have to bear is the fact that springs *cannot be scaled* – in fact, the formulas will tell you that once you study them.

The *third* point I would make is that the formulas appear to be much more formidable than they really are. Consider this:

> The safe load on a spring is equal to the stress in the wire multiplied by the area of the wire, divided by twice the diameter of the coil.

What could be simpler than that? Yet, this is precisely what expression (2) on page 6 is saying, no more and no less. So, when you come to any of these formulas just write – or speak – them out in words, instead of symbols. The symbolic forms make it much easier to put the actual figures in, and that is the only reason (other than saving page space, of course!) for using them!

Finally, on this question of formulas, none of them requires the use of anything but simple arithmetic. You can get your offspring to program them into the home computer (if they are willing to spare it!), and it can reel off spring after spring for you to try; you don't need a PDP11 to be able to allow for all the corrections, or even all the "preferred" proportions. In fact, in many cases, you can program a pocket calculator of the more sophisticated type to do the same. However, I think you will find that my charts will "break the back" of the work for you, and you will need the formulas only as a final check. But do make a test spring before you go into production in quantity. The formulas are not perfect!

I intend to start with the coil springs, both in tension and compression, and will work through a few examples. These will, I hope, show "how easy it is" as well as calling to your attention the limits of good and bad practice. Obviously, the spring has to be wound, and we shall look at various methods available, including a few "aids" which have been proved to be helpful. Leaf springs, both solid and laminated, come next, with some emphasis on the problem of the "scale spring". I have gone into the design of compound springs – using leaves of two different materials – in some detail, as well as alternative solutions to the "scale" problem. After covering some of the less usual types of spring, including coil springs used in torsion, I have taken a few pages to deal with cam return springs – such as I.C. engine valve-springs. Here again there are alternative graphical and mathematical methods available. I shall *not* be dealing with either clock drive springs nor with balance escapements (hairsprings) as these are very specialized – and in any case, even professional horologists usually buy these in ready made. For the same reason I have not covered such applications as spring washers, Belleville discs, and the like. Frankly, I think I have covered enough ground as it is!

CHAPTER 2

Tension and Compression Spring Principles

How do they work?
Look at Fig. 1. Here I have chopped off a piece of spring at "A" and "B", which has a load tending to stretch it. This means that there is a force, suggested by the arrows, tending to pull "A" upwards and "B" downwards. The effect of this force is to try to *twist* the spring wire at "C". There will be a little bending action in the lengths AC and BC, of course, and we shall deal with this later, but the main effect is to twist the spring wire. So, the criterion for the spring wire will be the *shear strength*.

In Fig. 2, I have cut the spring across at "C" and you will see that the twist, or "torque" in the spring wire is the load "W" multiplied by the lever arm, the length of which is D/2, D being the *mean* diameter of the coil. So, the torque is ½DW. Now, in a round rod subject to a twisting action, the stress is not uniform over the whole area of the wire, but is zero at the core and a maximum at the surface. I won't go into the mathematics of this, but those of you who like this sort of thing will already have worked it out in your heads. The maximum shear stress is given by:

$$f_s = \frac{16\,T}{\pi d^3}$$

But T = W.D/2, so that

$$f_s = \frac{8\,W.D}{\pi d^3} \qquad (1)$$

Or, to put it the other way round,

$$W = \frac{f_s\,\pi\,d^3}{8D} \qquad (2)$$

f_s = max. shear stress, lbf/sq.in.
T = Torque, lbf./in.
W = Load, lbf.
d = Wire dia. in.
D = Mean coil dia. in.
π = 22/7 for this sort of work.

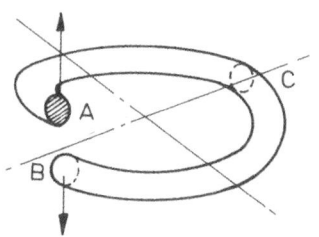

Fig. 1 *Showing how the load from adjacent coils applied at A and B exerts a twisting force at C.*

These are the basic equations for LOAD, and you will see that the load which a spring can carry is proportional to the stress, to the CUBE of the wire diameter, and inversely proportional to the coil diameter. So, doubling the wire diameter makes the spring 8 times as strong and doubling the coil diameter halves its load-carrying capacity, given the same material and working stress. Obviously, a "scale spring" is impossible! Another point; looking at Fig. 2 again you will see that the NUMBER of coils makes no difference to the load which a spring can carry. The second half coil shown there is under exactly the same load conditions as the first, and no matter how many coils (or half coils!) there are this applies. A ten-coil spring will carry exactly the same safe load as one with only two if the wire and coil diameters are the same. (And, of course, the material is identical.)

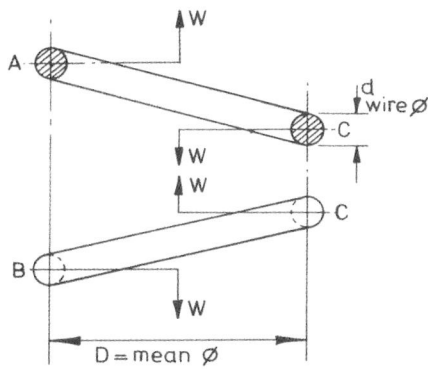

Fig. 2 *The twisting moment on the spring is ½D x W and is the same on all coils.*

Deflections

The number of coils does have an effect on the spring performance, as you all know; and in many cases, the spring deflection under load is almost as important as the load itself. Look at Fig. 3. Here I have shown two half-coils again, with a dotted outline showing the movement of the coil caused by the applied load. In the upper one, the end "C" has moved downwards by an amount "§". The corresponding end C of the bottom half coil is thus displaced downwards by this same amount. But

Fig. 3 *When the load is applied, C in the upper half-coil deflects by an amount §. This dislodges BC to bc. This also deflects under the load, so that the total deflection will be 2§ at B.*

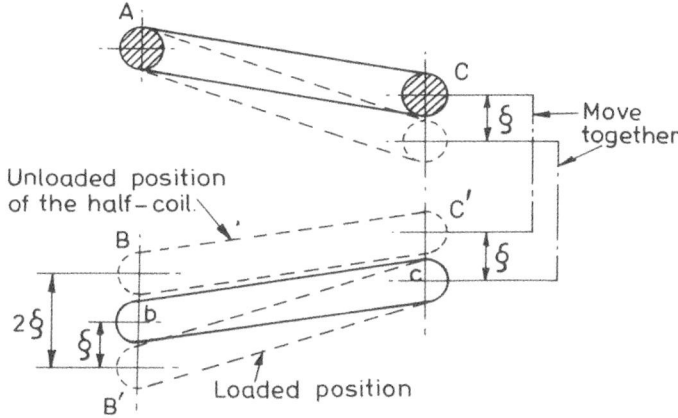

Tension and Compression Spring Principles 7

the end "B" of this bit of spring will, in addition, move a further "§" due to the load on it. Hence the total deflection is 2§. Add another pair of half coils and the next end will move by 4§, and so on. The total deflection depends *directly* on the number of coils.

In spring work, this deflection is usually stated as the RATE (or SCALE) of the spring, "R", and in Imperial measure is defined as the load in lbf to cause a deflection of one inch – lbf/in. (In S.I. units, this would be stated, for our size of spring, in Newton/mm.) The "math" of this *is* rather complicated (except to those home computer operators!), so I will simply state the formula this time. The deflection under a load W is given by:

$$§ = \frac{8W.D^3.n}{G.d^4} \quad (3)$$

§ = deflection, inches
n = number of active coils
G = Torsional modulus of elasticity. Lbf/sq.in.

This can be more conveniently written in terms of the "rate", W/§, thus:

$$\frac{W}{§} = R = \frac{G.d^4}{8D^3.n} \quad (4)$$

R = spring rate, lbf/in. of deflection.

(Note that I have given the units so far in the old "Imperial" – inches, lbf, and so on. There is no change in the formulas when consistent S.I. units – millimeters, Newtons, etc. – are used, and the charts provided later allow working in either system.)

From (3) above you will see that the deflection for a given load depends directly on the size of that load, on the cube of the coil diameter, on the number of coils, and inversely as the *fourth* power of the wire diameter. And, of course, inversely as the value of "G", which depends on the material used – AND on the temperature of that material, a fact not always remembered when dealing with springs working at (e.g.) the temperature of boiler steam. However, the main point is that so far as deflection and "rate" are concerned it is even more impossible to make a "scale spring".

I repeat, there is no need for the average model engineer to use these formulas at all, as you can use my charts, but looking at expression (2) and (3), if we want a spring which exerts a load of W lbf at a deflection of § inches and D diameter, we just write in the figures in (2) at a chosen stress to find the spring wire diameter d. Then put these figures into expression (3) to find the number of coils. There are two difficulties. The first is that we may (almost certainly will) find that this gives a ridiculous number of coils; or that the wire diameter won't allow the spring to be wound – you can't wind a $3/16$in. dia. spring with 12 s.w.g. wire! So, you will have to go back and try again, with either a different coil diameter or a different material. Eventually, you will come up with a spring that will serve, but it takes time – a lot of time. (That's why I started using charts, a long time ago!) But that is not the most serious snag. A spring designed that way may not give the right results – it is likely to take a permanent "set", for example. We must spend a minute or two on this, because it is important.

Correction Factors

In arriving at expression (1), we assumed that the torque was applied as in the case of a drive shaft. But a coil spring is

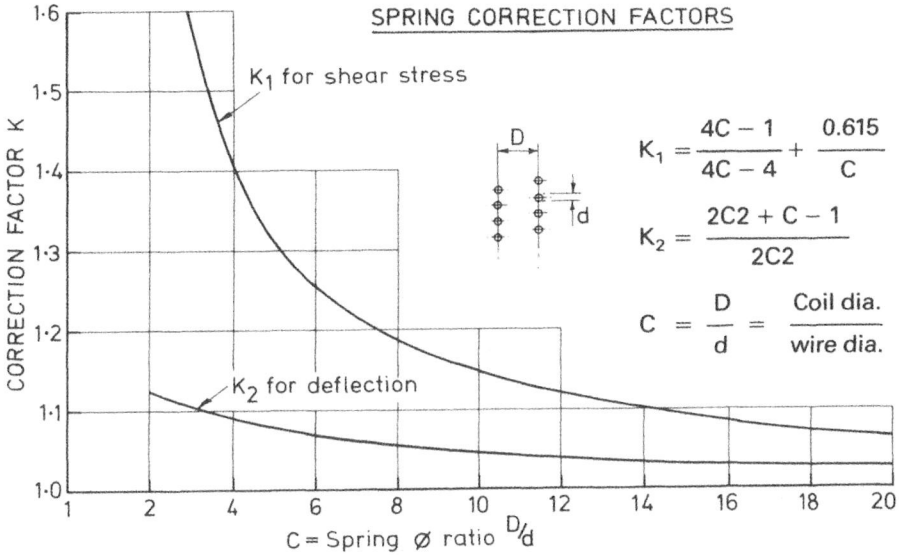

Fig. 4 *Curves of correction factors needed because the spring wire is not a straight bar. "C" is the ratio D/d. These corrections are allowed for in the calculation charts shown later.*

COILED, the shaft is curved. This makes a big difference, for the surface stress is no longer uniform around the circumference. It is higher on the inside of the coil than on the outside. And, while we are at it, we ought to allow for the small bending stress in the wire which I mentioned earlier. A little reflection will tell you that if the correction is needed because the wire is curved, it is likely to be greater the greater the curvature, and you would be right. It depends on the ratio of coil diameter to wire diameter. Look at Fig. 4. The upper curve shows the correction to be applied to expression (1), page 6. You will see that the curve rises very steeply as the ratio c = D/d decreases. Even with a relatively "easy" spring made from 16 s.w.g. wire on ⅜in. mean diameter, the actual stress will be 25% greater than that calculated, and would be 10% greater if it were ⅞in. mean diameter. The lower curve shows the effect on the deflection formula No. 3. These corrections, K_1 and K_2 can be derived mathematically, but those shown have been found from a very comprehensive set of experiments, and then reduced to curves, from which the mathematical models ("formulas") shown on the graph have been derived. So, the formulas you must get the kids to use on the computer are:

$$f_s = K_1 \frac{8WD}{\pi d^3} \qquad (5)$$

and

$$R = K_2 \frac{Gd^4}{8D^3 n} \qquad (6)$$

with K_1 and K_2 either read from the graph or calculated from the value of D/d. But – see how complicated it has become all of a sudden. Indeed, we are in the same position as the designer of a bridge. The largest load a bridge has to carry is its own weight. But you can't determine the weight until you have designed the bridge. And you can't design . . .! This is what I am reliably informed is a "catch 22 question". (Actually, it is more like catch 2222, for there is wind-loading and so on as well!) Now go out and see how many bridges you can find – it is getting over that sort of difficulty that makes "Engineering" the art it is. Similarly with spring design – there are ways out.

The first is to use charts, as I have done for years. If these are drawn out, using all the corrections, you can see the effect of altering this or that value at a glance. Or you can do it with "chips" – a home computer or a programmable calculator. But there is another help, too. Over the years, so many springs have been used that we have found out that there are values of D/d which are "good", others which are "satisfactory" – and so on up to those which are plain nonsense. I have already mentioned a 12 s.w.g. spring $3/16$in. mean diameter – the mandrel would be only $1/16$in. dia. – less, for there is "spring-back" to allow for. Now imagine one made from 6 mil. piano wire 2in. coil diameter – D/d = 330! I will give you details of the ranges later, but can say now that while springs CAN have a wide range of ratios, the majority lie between D/d = 5 and 14; a few very "soft" springs may lie between 14 and 25, and very few indeed "harder" than D/d = 4. These limits will be shown on the charts.

Though this introduction may leave you with the idea that the job is difficult, that isn't true. It does take *time* – even with a full computer program and years of experience, the commercial spring-makers have to take care when designing a special spring – and nearly all model engineers' springs *are* special, simply because scale effects take us out of the "normal practice". But with the use of charts and nomograms I have, over the years, found that it is possible to reach an acceptable design after only a reasonable amount of work.

Spring Wire

Whatever calculation method is used, the designer does have to make a decision as to what stress to use, and this depends on the material. And that, in turn, means "what we can get". Some desirable types are just not available in reasonable quantities and others need heat treatment after winding. Most of "our" springs are wound cold, from hard wire. That hardness is achieved by repeated drawing through dies, and each drawing operation increases both the shear and tensile strength – it is work hardened. So, the shear stress we can use does depend on the wire size – in general, the smaller the diameter the higher the allowable stress. This can be surprisingly high to those used to normal working stresses – the torsional YIELD point – elastic limit – of 10 mil. piano wire, for example, is nearly 200,000lbf/sq.in. – about 1400 Newton/sq.mm – and falls only to 140,000lbf/sq.in. at 0.080in. dia. For most duties, we have to keep below this elastic limit, and the usual rule for compression springs is that when it is compressed so that all coils are touching, the stress should lie just below this elastic limit.

The "regular" spring wire we get is what is known as *"Patented carbon steel spring wire". The* word "patented" does not refer to the patent office, but to a process applied to the steel to make it easier to draw into smaller gauges. It is typically between 0.65 and 0.75% carbon with perhaps 0.75% of manganese but no alloying content. It is available in common wire gauge sizes.

Also fairly readily available is *"Piano"* or *"Music"* wire which, as its name implies, is intended for use in stringed instruments. It has a higher carbon content – 0.85-0.95% – to give a higher tensile strength, for, in pianos especially, the wires are very tightly stretched. The shear strength is correspondingly increased. It has the advantage (apart from higher permissible stresses) that it is available in many more sizes which come in between the s.w.g. diameters, but the disadvantage is that the higher tensile strength makes it rather more difficult to wind. Both of these carbon steel wires can be had either zinc or cadmium coated (*not* electroplated) *before* the final drawing process – the wire is drawn through the dies after coating. This not only improves corrosion resistance but also improves the fatigue performance; the soft-metal coating reduces the surface roughness which may arise when drawing, and from what I have said already you will appreciate that *any* surface defect on a coil spring wire is highly undesirable. (Even a thin coating of rust!)

18-8 hard-drawn stainless steel is a very useful material, especially for situations where the temperature may be high or there is risk of corrosion – it can be worked up to 300 deg. C, whereas carbon steel wire is a little unhappy above 125 deg. C. It has an elastic limit in shear very slightly higher than regular carbon steel. (The working stress must, of course, be reduced when applied in hot environments, as I have already remarked.) This material is expensive and not too easy to wind. *Hard drawn phosphor bronze* also is non-corrosive so far as steam/water is concerned, but is normally recommended for continuous use only below about 110 deg. C. It is relatively easy to obtain in a wide range of gauges or to metric dimensions.

There are other materials, but few are suitable for use by model engineers. The prime spring material, *chrome vanadium steel,* for example, must be heat treated after winding, as must *beryllium-copper* – i.e. the spring is wound in the "soft" condition and then hardened. *Monel* can be wound hard drawn and will safety withstand both sea-water and corrosion and temperatures up to 225 deg. C. The shear elastic limit is about the same as phosphor bronze. Finally, *hard-drawn 70/30 brass* is a very cheap spring material for cases where mild conditions apply. It has strength properties about two-thirds of phosphor bronze and should not be used above 80 deg. C but, oddly enough, can be used at low temperatures. The most usual application of "spring brass", however, is for flat springs, especially where sharp bends may be needed.

Table 1

Material	G. Lbf/ sq.in. × 1,000,000	G. Newton/ sq.mm × 1,000
Carbon Steel	11.4	90
Piano Wire	12.0	83
18/8 Stainless	10.0	69
Phos. Bronze	6.0	41
70/30 H.D. Brass	5.0	35
Monel	9.5	65

Before going on to look at working stresses, Table I (above) gives the Torsional Modulus of Elasticity, G, for some materials.

Working Stresses

These are given in Figs. 5, 6, 7, and 8 for the four most usual materials. In each case, the upper curve gives the *minimum* torsional elastic limit. It would be rare for the material to lie below this figure, and "accidental" stresses, e.g. compressing coil-to-coil, or the stretching of a tension spring to install it, up to this limit can be tolerated. Even where such "accidents" can be designed out, of course, it is best to work 5% below the curve. Below this one is a curve of *"Working Stress for average duty"*. This is the stress which should be used in the charts for all normal applications – axlebox springs, for example – and is the stress at the maximum duty load. This is calculated arbitrarily at 70% of the elastic limit for carbon steel and bronze, and 80% for stainless steel and music wire.

Please note that you can alter this to suit your own judgement; a spring which has to do no more than sit there and exert a force can be worked higher

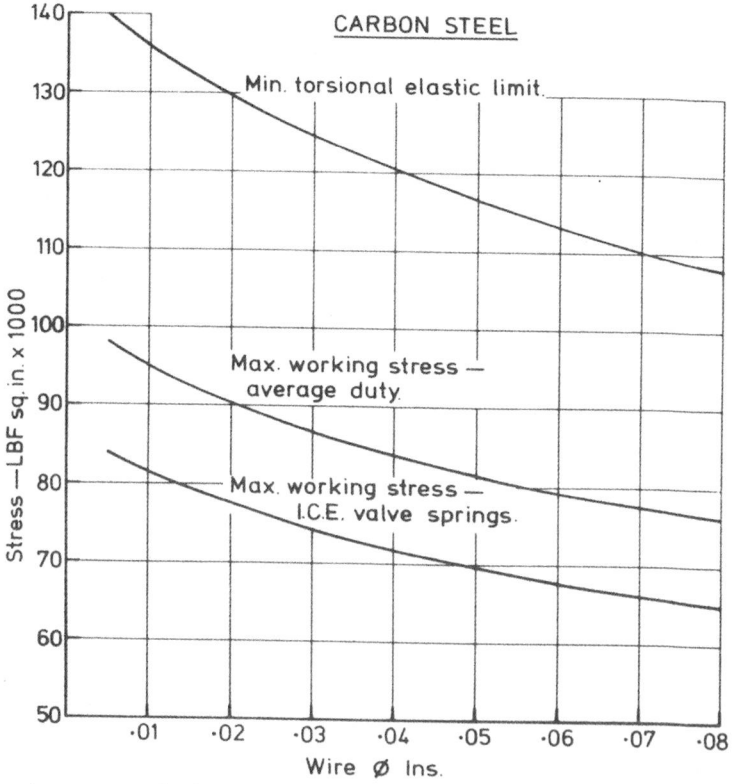

Fig. 5 The stress curves for "patented" carbon steel ("Regular") spring wire.

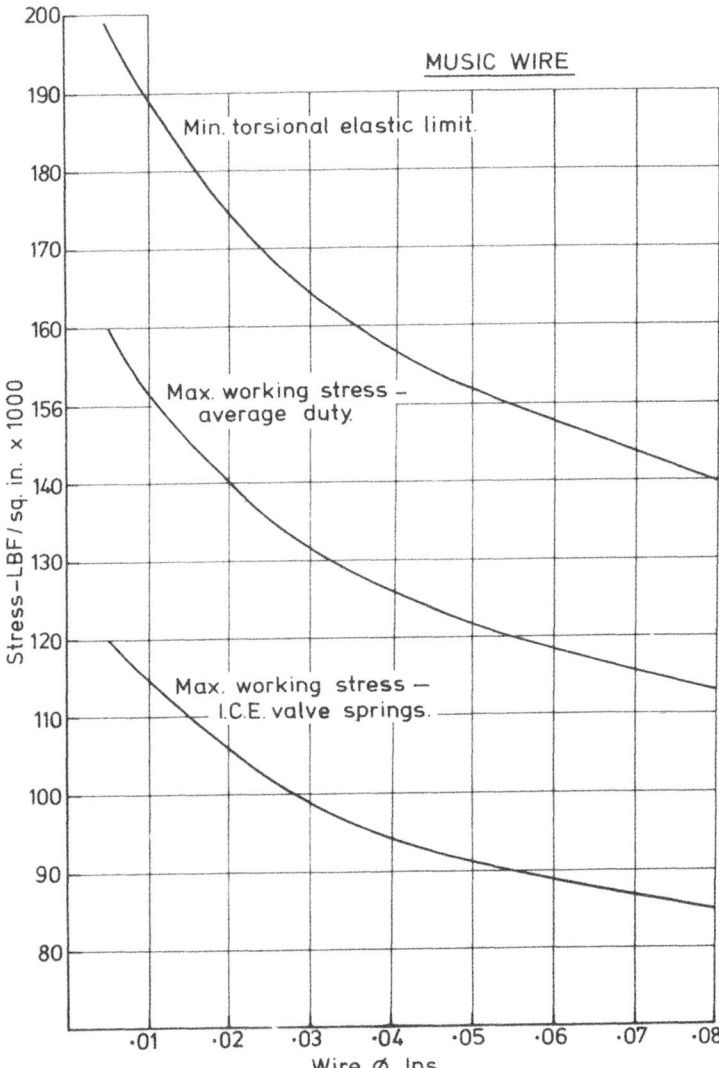

Fig. 6 *The stress curves for "Music" or "Piano" wire.*

as long as it can never exceed the elastic limit. One that carries a dead weight with some little movement –an axlebox spring again – needs the 70% assessment; on the other hand, one that has a light load but is repeatedly stretched (a lubricator ratchet spring, perhaps) might be better designed at a lower figure. *There is no*

Tension and Compression Spring Principles   13

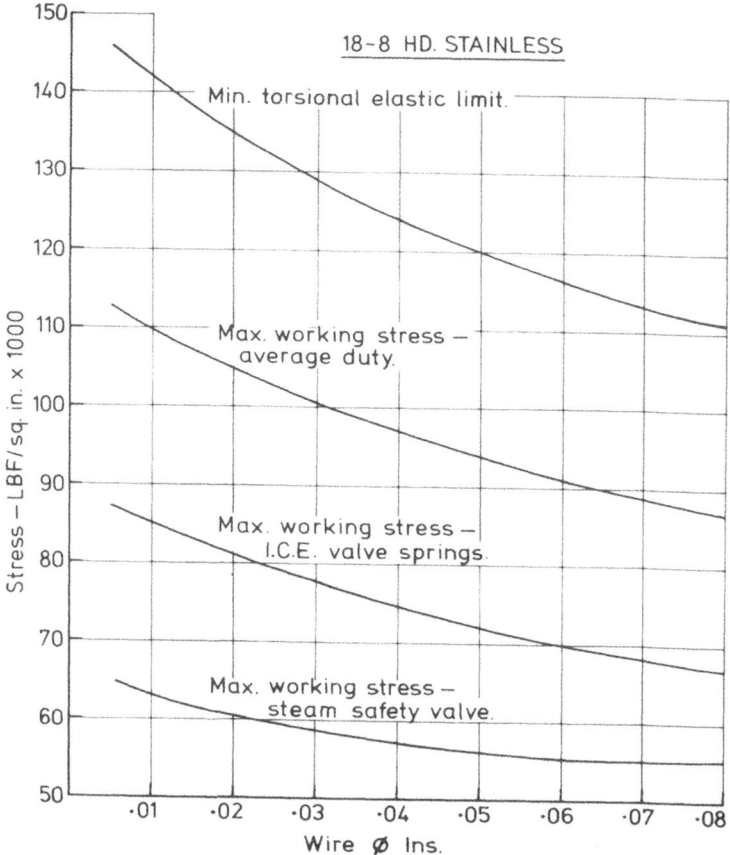

Fig. 7 *Stress curves for spring quality drawn phosphor bronze wire*

harm in using a low stress – if you don't mind using heavier wire and more of it. The one exception to this is, perhaps, the case of the I.C. engine valve spring, which I will be dealing with later.

This is a classic case of the "Severe Duty" curve, which appears on the graphs. The severity is not so much due to fatigue as to the fact that the inertia of the valve spring itself may mean that the whole load is carried on just the top few coils at the beginning of each lift.

In addition, such springs may vibrate axially from coil to coil, again increasing the local load. But, for reasons which we shall see when I deal with this class of spring, we always try to work UP to this stress, rather than go below it.

Finally, on the Stainless and the Bronze charts you will see a curve for safety-valve springs. These take account of the change in properties at the steam temperature – at 100lbf/sq.in. for the stainless steel and at about

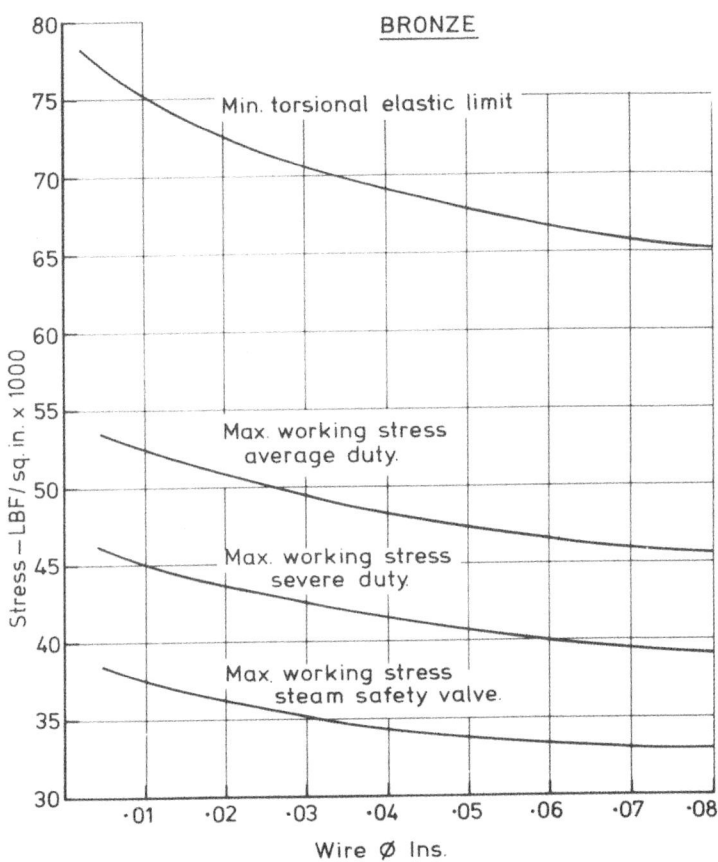

Fig. 8 *Stress curves for hard drawn (spring quality) 18-8 stainless steel.*

30lbf/sq.in. for phosphor bronze – one normally would not use this material at high pressure. Incidentally, if you CAN get zinc or cadmium-coated steel wire, these can be used on steam safety valves up to about 50lbf/sq.in., at about 50% of the elastic limit, as model steam boilers are not normally used for long periods at a time. (Coated before drawing, that is, *not* electro-plated.) You will see that these charts give figures only up to 0.08in. wire diameter (14 s.w.g.). You should not need wire much larger than that, but if you do, the curves may be projected backwards as far as 0.1 in. (for example, to No. 12 gauge) in a straight line with little error.

In the last analysis, the choice of material usually lies in "what you have on the shelf". It just isn't worthwhile ordering (and waiting for delivery of) an alternative size of wire for a run-of-the-mill spring. In the books, you will find rules for an "efficient" spring, such as:

Tension and Compression Spring Principles 15

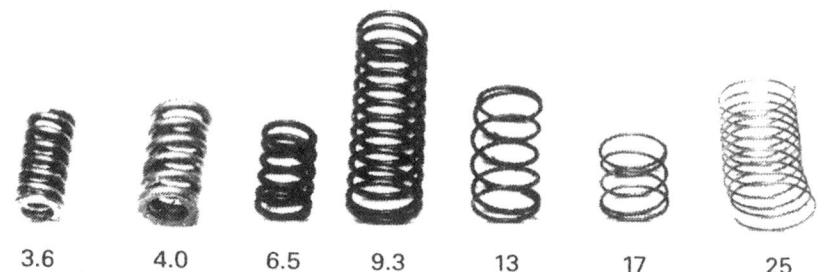

Below. Fig. 9 *A range of springs wound to increasing D/d ratio. All except that on the right are commercial springs.*

D/d between 7 and 9; free length of compression springs to lie between 2 and 4 times the mean diameter; distance between coils at a maximum load to be 10% of the wire diameter (again, for a compression spring), and so on. This is important if you are using 1,000 springs a day–the saving in the cost of wire over a year is considerable. But even in industry, many springs depart widely from the "ideal". So long as the spring is, like the engine power of a Rolls-Royce, "adequate for its purpose", and so long as you keep a sense of proportion, there is no need to be too worried about "ideal" conditions.

To help you assess these things, look at Fig. 9. This shows a series of compression springs of varying D/d ratios. All except that on the right are from working mechanical devices and gave satisfactory service; I wound the one with D/d = 25 just to show you what it would look like; rather unstable. In Fig. 10, I show three lengths of stock spring from 4.5 to 10.7 ratio, all of the same wire diameter. Fig. 11 illustrates the point about length/diameter ratio. The upper

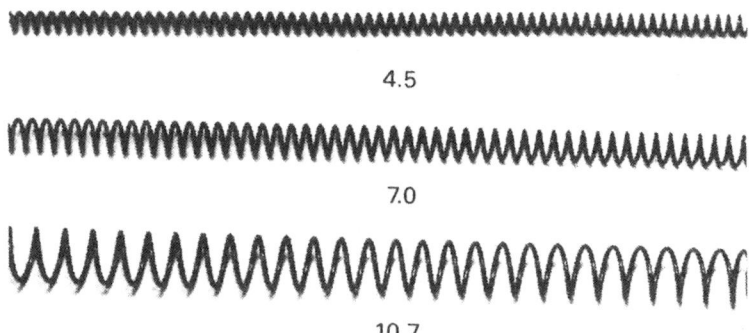

Fig. 10 *Three lengths of "stock spring" of the same wire diameter but made with differing D/d ratios. The D/d ratios are inset on the photograph.*

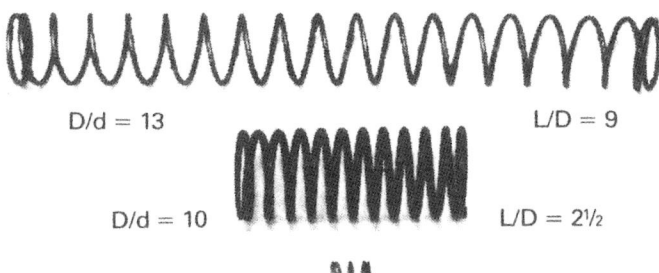

Fig. 11 Springs of similar D/d ratio, but varying length/coil dia. proportions. All three are from "important" mechanisms, but the longest one was guided laterally.

spring is very long and with a moderately high D/d figure. It is the return spring from a pneumatic actuating device on an aircraft. In service, it sat inside the cylinder and was guided against bucking. Below is a valve-spring from a 60-year-old 20HP Rolls-Royce, while at the bottom is a short spring from a diaphragm-type pump. All three are "important" springs, yet show a fair range of both D/d and L/D ratios. Fig. 12 shows two springs of the same length/diameter ratio, one compression, the other tension. The latter will present no problems at all, but the compression spring will certainly buckle unless it is guided. Finally, Fig. 13 is a tension spring having D/d fairly high at 16, and a relatively short length. You can see how the designer has got over the problems of the securing eyes by winding the ends to a conical shape.

Fig. 12 Two springs of similar length/diameter proportions. The tension spring is normal, but the lower one would need a guide.

Below, Fig. 13 A tension spring of large D/d ratio. Note the conical ends to transfer the load to the hooks.

Tension and Compression Spring Principles   17

CHAPTER 3

Compression and Tension Spring Design

Coil Spring Design Charts

For convenience, I show in Table II the dimensions of both the "Standard Wire Gauge" and the "Music Wire Gauge" in inches. (Note that the *American* Music Wire Gauge is slightly different.) "Piano" and "Music" wire may be taken as identical.

Now look at Figs. 14 and 15. These are really one diagram split into two, as I have tried to cover a very wide range of spring loads – from 1/10lbf up to 100 – and have split the diagram at W = 3lbf. These charts have been prepared using Imperial units, inches and pounds force, and are based on the Imperial Standard Wire Gauge, but there are secondary scales of millimeters and Newtons for younger readers. The vertical scale is that for MEAN COIL DIAMETER, and the horizontal gives "Maximum Working Load" – which is, the highest normal load the spring will be required to carry. This is based on stresses given in Fig. 5 for "Average Duty" for carbon steel spring wire. (How to use the diagram for other materials or stresses appears later.)

The chart also includes an allowance for the correction factors K of Fig. 4. Running downwards from left to right is a series of lines each marked with a wire gauge – shown at "even" gauge numbers for the most part, but don't worry about that for now.

Suppose we want "a" spring which will carry a load of 1lbf and there are no other constraints. Fig. 14, read up from 1.0; we can use 28 s.w.g. wire with a coil 0.115in. dia.; no. 26 s.w.g. with coil 0.22in. dia.; 24 s.w.g. with a coil 0.41in., or 22 s.w.g. with a coil 0.8in. dia. What a choice! But, there is another set of lines running down from right to left, giving the D/d ratios. Our first choice of 28's wire lies almost exactly on D/d = 8, and from what we have said earlier this is a "good" ratio. 26's lies at about D/d = 12, not at all bad, but 24's will be at something like 19, not so good; and finally, 22's gauge wire lies in the dotted region, where (unless the spring is guided or restrained) it will certainly tend to buckle under the load. So, in this example we should use 28's gauge wire wound to give a mean coil diameter of 0.115in. diameter. The internal diameter will be 0.115in. – 0.0145in. (the wire dia.), which is 0.1 in. as near as makes no odds. (We shall see later that the mandrel diameter would have to be smaller than this, to allow for spring-back when winding.)

Table II S.W.Z. & Music (Piano) Wire Sizes

Gauge number	S.W.G. Inches	Music Inches	Gauge number	S.W.G. Inches	Music Inches
6/0	0.464	0.0065	22	0.028	0.048
5/0	0.432	0.007	23	0.024	0.051
4/0	0.40	0.0075	24	0.022	0.055
3/0	0.372	0.008	25	0.020	0.059
2/0	0.348	0.0085	26	0.018	0.063
0	0.342	0.009	27	0.0164	0.067
1	0.30	0.010	28	0.0148	0.071
2	0.276	0.011	29	0.0136	0.074
3	0.252	0.012	30	0.0124	0.078
4	0.232	0.013	31	0.0116	0.082
5	0.212	0.014	32	0.0108	0.086
6	0.192	0.016	33	0.010	0.090
7	0.176	0.018	34	0.0092	0.094
8	0.160	0.020	35	0.0084	0.098
9	0.144	0.022	36	0.0076	0.102
10	0.128	0.024	37	0.0068	0.106
11	0.116	0.026	38	0.006	0.112
12	0.104	0.028	39	0.0052	0.118
13	0.092	0.030	40	0.0048	0.125
14	0.080	0.032			
15	0.072	0.034			
16	0.064	0.036			
17	0.056	0.038			
18	0.048	0.040			
19	0.040	0.042			
20	0.038	0.044			
21	0.032	0.046			

Standard Wire Gauge continues to No. 50 at 0.001in. dia., and Music (piano) wire to No. 45 at 0.160in. dia.

Metric wire sizes are not yet standardized, and will, in most cases, correspond to metric equivalents of the above, rounded off to 0.002mm.

We will try another example, more difficult, in a minute, but first take another look at both charts. You will see that these D/d lines are "coded", "8" is full black, representing the so-called "ideal" ratio. The lines for 5 and 14 are chain dotted. A spring lying between these two lines can be regarded as acceptable for almost any duty. The lines for 3 and 25 are plain dotted, and the area below 3 and above 25 is a "caution" region. Springs between D/d = 5 and 3 CAN be used, but they will be very stiff and those between 14 and 25 will be "soft". Below 3 you will find it almost impossible to wind the wire. Above 25 the coils will tend to disorder even if guided. You *can*, however, use ratios between 3 and 5, or between 14 and 25 if there is *no other way out*. Winding a

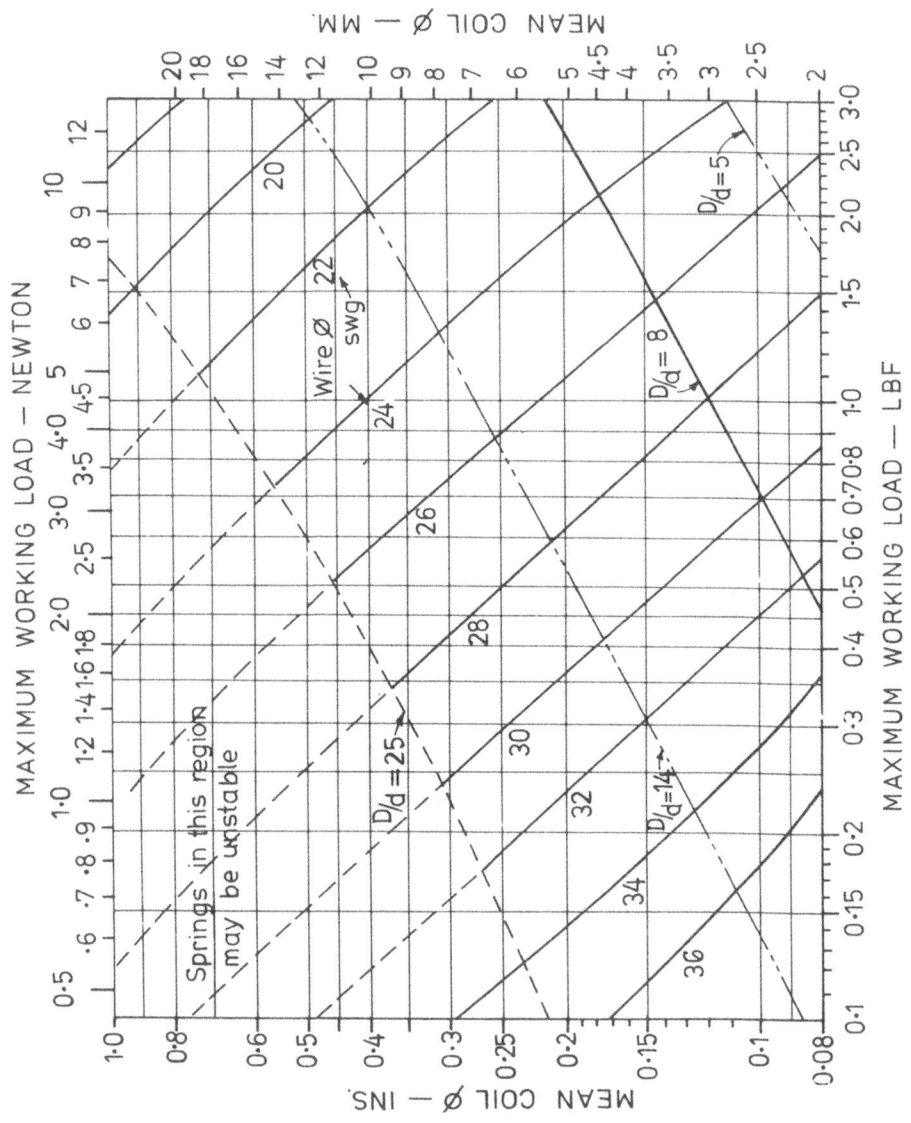

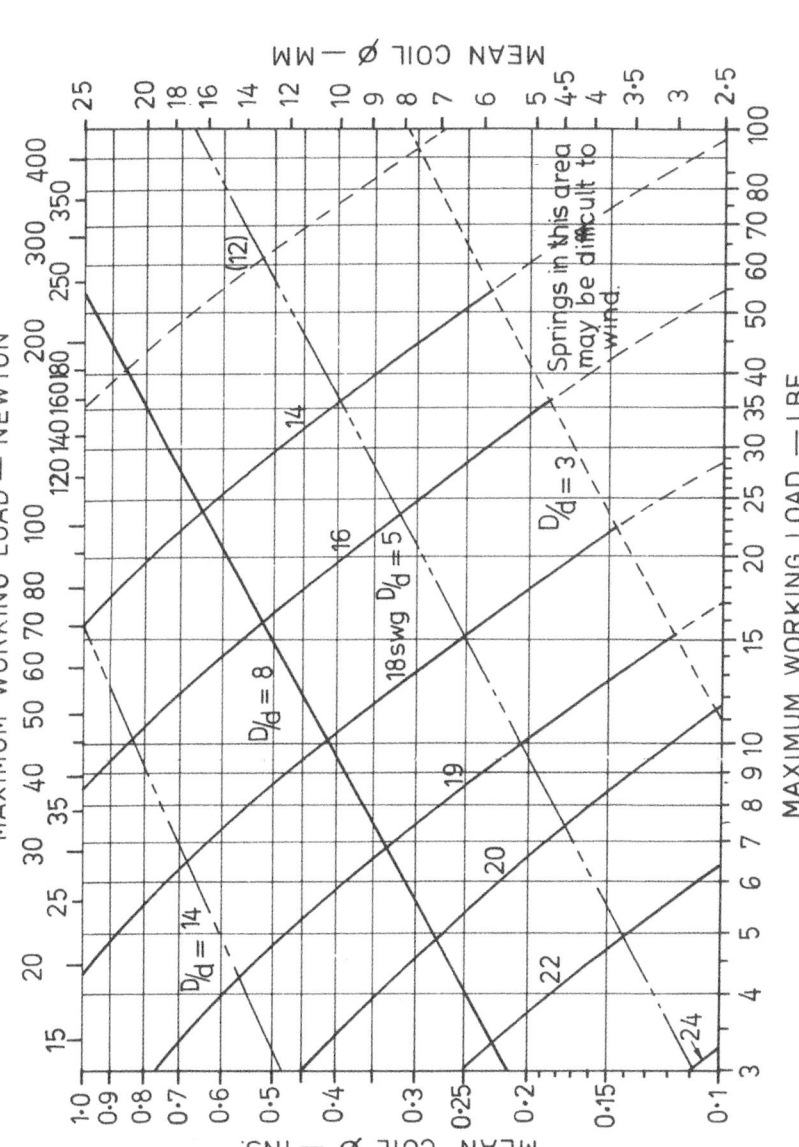

Fig. 14 (opposite page) and **Fig. 15** (below) show curves of approximate working load at various wire and coil diameters. The loads are calculated on a shear stress, f_w, equal to 70% of the torsional elastic limit for "regular" carbon steel wire at each diameter (Fig. 5) and include the allowance K1 shown in Fig. 4. To use for any other stress, f_s, either DIVIDE the required load by f_s/f_w, or multiply the figures on the load scale by f_w/f_s. See text.

Compression and Tension Spring Design 21

spring for D/d = 4 is not easy, but can be done. Springs lying between 14 and 25 may be a bit more fragile than usual, but are practicable if you take care. In other words, when using the chart, keep within the bands 5 and 14 if you possibly can, and the nearer to D/d = 8 the better if there is a choice.

Now let us try another example – Fig. 15. We need a spring to exert a force of 9lbf, which will fit inside a ⁵/16in. dia. tube. The "mean" diameter will be less than 0.3125, so look along the 0.3in. line until it meets the 9 lbf vertical. Where they cross lies between 19 s.w.g. and 18. Which to use? Remember, this chart has been drawn for an arbitrary stress of 70% of the elastic limit, and there is no reason why we shouldn't use a lower or, if it is safe, a higher figure. 19's gauge at a slightly higher stress (DOWN the vertical line) or 18's at a lower (UP the line) will fit the bill. You have 19's gauge in stock, but no 18, so use that. 19's is 0.040in. dia., so to pass through ⁵/16in. the maximum mean diameter will be 0.3125–0.040 = 0.272in. So, we must try again, and the coils of the spring will expand radially as it is compressed try 0.26in. this time. This brings us nearer to the 19's gauge line, and we must use another chart – in a minute or two – to check the stress. The main purpose of Figs. 14 and 15 is to give you an "inspired guess" at the size of the spring before you start the definite calculations, and in a few cases, may give you an answer you can use without further ado.

You will see in the caption to the chart that it can be used for materials other than the regular carbon spring steel. If you look at Figs. 5 and 6, you will see that 0.04in. dia. piano wire can work at a stress of 126,000lbf/sq.in. while regular spring wire will stand only 85,000. So, to solve the last problem for music wire, all you have to do is to DIVIDE the load by 126/95 = 1.33. Then, use the chart as before. 9/1.33 = 6.77lbf gives us a choice between 19's and 20's gauge wire. In the previous example, for a 1lbf spring, ¹/1.33 = 0.75 and (Fig. 14) a 30's gauge spring might do – we can check the stress from the next chart in a moment. So, to use Figs. 14 and 15 for other materials, all you need to do is to multiply the load scale by the factors given in the captions and then use the diagram as before.

Stress Checking

We can now turn to Fig. 16 and refine the design to more exact figures. This is a NOMOGRAM – a device which enables us to see the effect of any of four variables on the solution of a problem. The upright scales are logarithmic and work in pairs – A_1 on the left with A_2 on the right, and similarly and B_2. Take the example we have just been looking at. We have a load of 9lbf (scale B_2) and, with 19's gauge wire and 0.26in. mean dia., D/d = 0.26/0.04 = 6.5 on B_1 Lay your ruler across these two uprights B_1 and B_2 and, where it crosses the center vertical make a light pencil mark, as I have indicated. (Use the correct scale – it is easy to forget and draw the line from 9 NEWTON by mistake!) Now, from where this line crosses the center draw another from 19's gauge (0.04in.) on A_2 across to the stress scale on A_1 it shows about 92,000lbf/sq.in. Check from Fig. 5, where we see that the elastic limit for 0.04in. wire is 120,000, so that the working stress is 92/120 = 0.76 of the elastic limit – safe. Now try the example where we want to use piano wire to give a spring to carry a load of 1 lbf. A few minutes ago we saw from Fig. 14 that the factored load should be 0.75lbf and what a spring of 30's gauge wire might do on

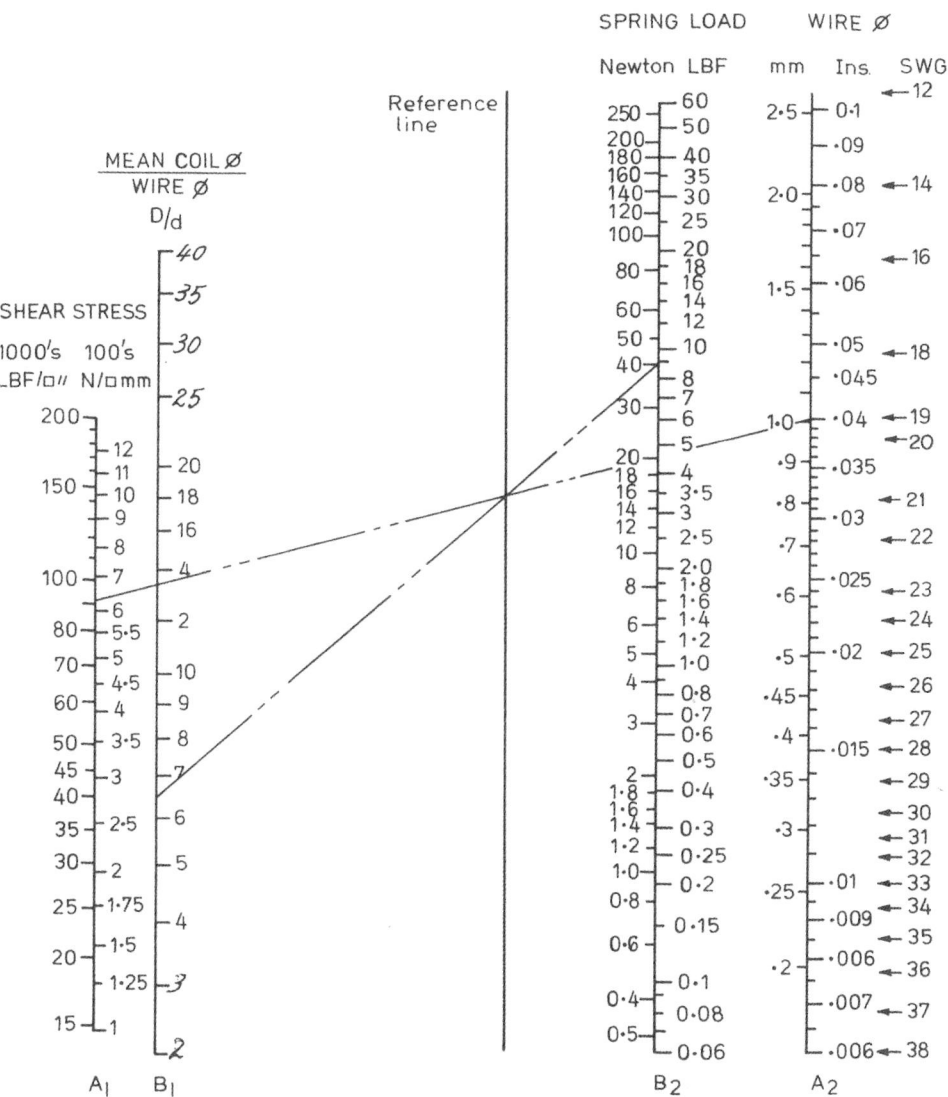

Fig. 16 *Nomogram connecting wire diameter, d; D/d ratio; spring load; and the actual stress in the wire. Allowance for factor K_2, Fig. 14 is included. This chart can be used for any material. The chart has been prepared in Imperial units; S.I. (metric) figures have been obtained by "conversion".*

a mean dia. of 0.095in. coil diameter. Hence D/d = 095/0.0124 = 7.7. Try this on Fig. 16 BUT USE THE PROPER LOAD OF 1lbf. Try it yourself and then check that, from Fig. 6, the stress is only 64.5% of the elastic limit for 0.0124 wire. Easy, isn't it?

Now suppose you have some 20's gauge bronze wire and want to make a small safety valve spring, which must pass over a $3/_{16}$in. dia. rod and exert a force of 1.7lbf. The wire is 0.033in. dia. so, with a small clearance, the mean diameter will be about 0.226in., from which D/d = 0.033/0.226 = 6.85. Lay your rule across the "B" columns and mark the reference line where it crosses. From the wire diameter on A_2 lay your rule again across this mark. It crosses the stress line, A_1, at about 29,000lbf/sq.in. Reference to Fig. 8 shows that this is well below the suggested working stress for bronze wire 0.033in. dia. when used at steam temperatures. The spring could be used at 40lbf/sq.in. and possibly higher for very intermittent use.

Again, it often happens that you *have* a spring already but are not sure what load is safe to apply. Take the case of the large tension spring in Fig. 13. The larger coils are 2in. mean diameter and the wire 0.128in. dia., giving D/d = 15.6. The safe working stress for average duty is (Fig. 5) off the chart, but projecting the lines linearly would be about 75,000lbf/sq.in. Lay a rule from A_1 to A_2 and mark the reference line. Then, from D/d on B_1 and project across to the load line, B_2. I get a figure of 30lbf. We can, in the same way, check the absolute maximum load at the elastic limit, which by my estimate would be 180,000lbf/sq.in. Try it yourself – I get a limiting load of 42lbf. We should to check the smallest coil as well, which is 0.8in. diameter. Try it yourself, working out the D/d ratio and then the stresses. You will – or should – find that these small coils can carry an even greater load. You would, in a case like this, also check that the straight extensions of the wire will carry the load in tension; these are in fact safe, with a stress (load/area) of only 2480lbf/sq.in. at the working load.

You will find that this nomogram can be used in other ways, too. Given the wire diameter, for example, and the stress you mark the reference line. You can then very quickly try the effect of various coil diameters on the safe load – all you need to CALCULATE each time is the D/d ratio. Other applications will occur to you.

You must, of course, remember previous work on the D/d values and I have shown the "extreme" values in italic figures to remind you. There is just one point I should emphasize. The figures for s.w.g. numbers lie away from the vertical A_2. The little arrow against each is to remind you to project this point *horizontally* until it actually meets the vertical line.

As I said, the scales are logarithmic, and if for any reason you need to extend any of the scales, this can be done if you remember (e.g.) that the vertical distance between 20 and 25 on the load line will be same as that between 200 and 250; between 4 and 5 D/d will be same as that between 40 and 50; if you want to go up to 100lbf on B_2 plot the distances between 6, 7, 8, 9, 10 above the 60 and call them 70, 80, 90, and 100; and similarly you can go from 0.1 to 0.15 by replotting the 0.01 to 0.015 spaces.

The S.I. Metric scales are, of course, "derived" – I plotted the nomogram in Imperial units – so if you want to extend these I do suggest that you do so in inches and lbf and convert. Incidentally, whereas Figs. 14 and 15 have the curves "smoothed", especially those for D/d (it is, after all, intended only to facilitate your "first guess" at wire diameter), the nomograms prepared with as high a degree of accuracy as an 8-digit calculator will allow. And though slight inaccuracies may creep in due to paper stretch and in the printing process they should, as reproduced, give results which are more than adequate. Errors due to tolerances on wire diameter or in your mandrel diameter are likely to be far more important.

Nomogram for Spring Rate and number of coils

As before, we have two pairs of related vertical scales with a reference line in between. A_1 is the "Rate" scale and A_2 shows the number of coils. B_1 is the D/d ratio we have used before, and B_2 shows wire diameters. You must be fairly careful when using this one, for two reasons. First, the two scales on the right are very close together, and it is easy to read the wrong one. Second, and more prone to cause error, the B_1 scale for D/d increases from top to bottom – *it is upside down* in relation to the others. In passing, I always use a celluloid square or a perspex rule on nomograms, so that I can see the figures above and below the line.

The chart is used in the same way as the previous one. Lay your rule from the known wire diameter to the known D/d ratio and mark the reference line. Then, from the chosen figure on A_1 rule across the mark on the reference line and read off the required number of coils on A_2. Or, of course, you can use it in reverse if you have a spring of known dimensions and want to know what the rate is likely to be. (Some springs are too stiff, or too "floppy" to measure this directly!) As we shall find out soon, it is, fairly often, difficult to get the spring rate that we need, and we have to alter the spring design right from the start. However, when using charts, this is easy, and while it might take a number of trials, just using the formulas we saw at the beginning the second trial with charts usually comes out as we wish. Apart from anything else, the chart enables us to see the effect of altering various dimensions just by looking at it.

As an example, look at the spring we used in the try-out of Fig. 14. You will recall that this was to exert a force of 1lbf, and we decided to use 28 SWG wire (= 0.0148in) on 0.115in. mean diameter, giving a D/d ratio of 7.8. As a first try, assume a spring whose length will be three times the diameter – for example, 0.35in. long when free. This gives room for 11 coils if there is a space between coils equal to the wire diameter. However, the coils at each end will be ground flat and will be inoperative, so that there are 9 *working* coils. See Fig. 18a. Apply the nomogram for these known figures, and you should find that the rate will be 4.9lbf/in. (Don't forget that the D/d scale is inverted.) To get a force of 1lbf, therefore, we need a compression of 1/4.9 = 0.204in. – for example, 0.2 as we can't work to one-thousandths! Then the compressed length will be 0.35 – 0.2 = 0.146in.

Here is the first snag; clearly this will not give enough room for nine free and two dead (total 11) coils of 0.0148 wire. To get them in, we must increase the overall length (the "height") of the spring. This will not make any difference to the rate – all you have to do is to wind the spring for a coarser pitch – and, of course, make sure that there is room for the longer spring on the actual job. But we don't want the spring longer than is necessary, so choose an arbitrary clear space between the coils when the spring is

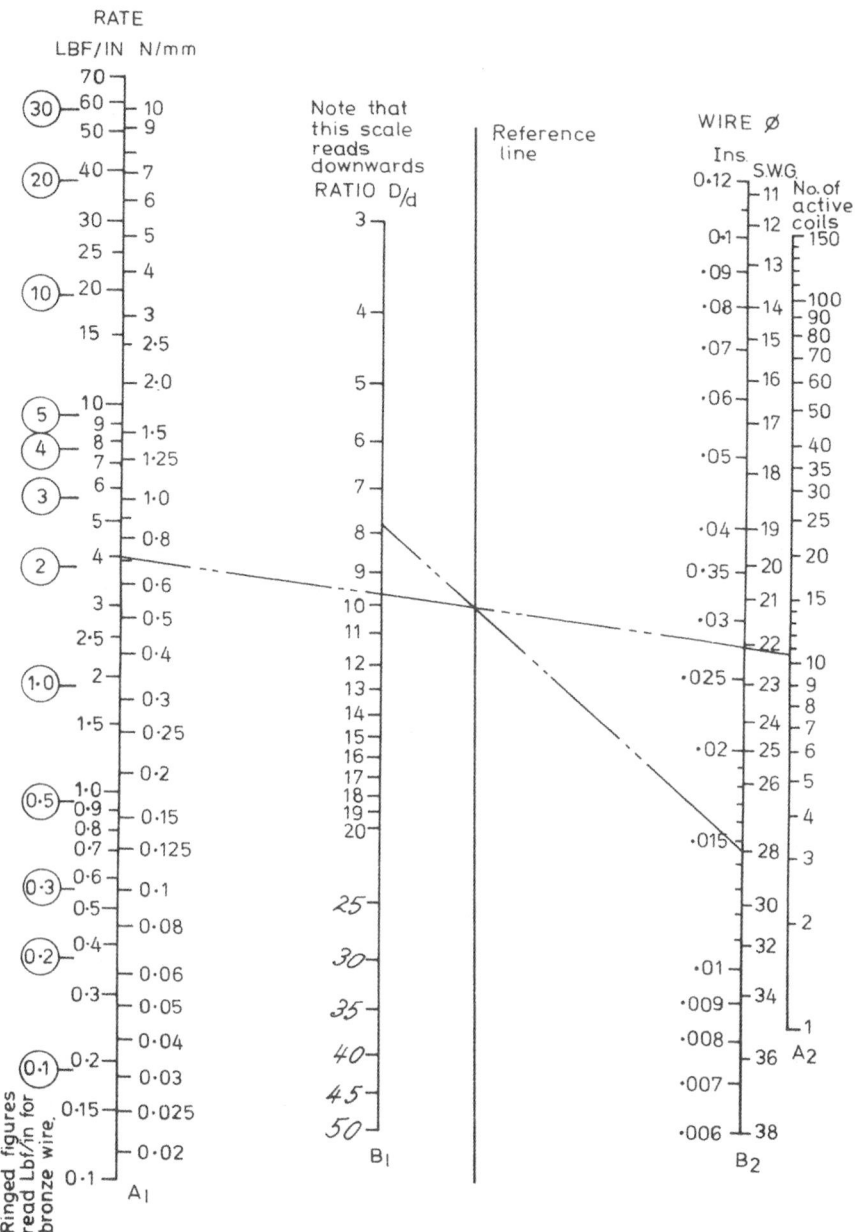

Fig. 18
(a) Spring as first designed.
(b) Revised design, to permit full compression.
Note: (a) is the "representational" and (b) the "conventional" method of drawing a spring.

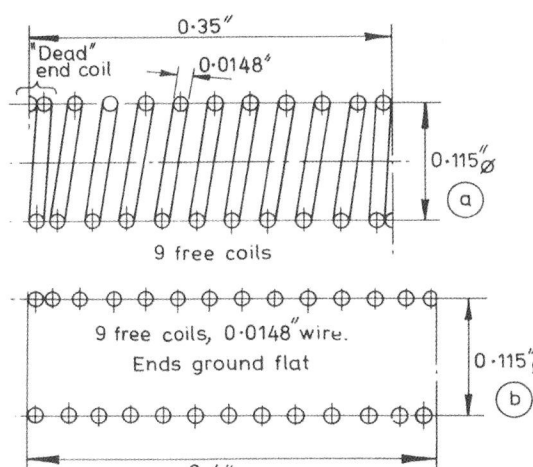

compressed. About 20% of the wire diameter, though less would probably do.

There are ten spaces between 11 coils, so the "load length" of the spring will be 11 × 0.0148 + 10 × 0.2 × 0.0148 = 0.192in. For example, 0.2 for convenience. So, the "free length" of the spring will be 0.4in. (0.2 compression plus the load length), which gives a ratio of length/coil diameter of 3½ – very reasonable. The final spring design is as shown at (b) in Fig. 18.

Now to finalize the job, check that when fully compressed the spring is not overstressed. The fully compressed length will be 11 × 0.0148 = 0.163in. Then 0.4 – 0.163 = 0.237, and this is the compression needed to bring the coils into full contact. The spring force in this condition = rate × compression = 4.9 × 0.237 = 1.16lbf. Go back to the nomogram in Fig. 16 and work out the stress at this load. (Yes – do it yourself; good practice!) You should find that it comes out at about 100,000lbf/sq.in., well below the elastic limit for that size of wire (see Fig. 5). Thus even if a visitor to your workshop picks it up and squeezes it flat "to see how springy it is" (they WILL do these things!), no harm is done. So, you have designed a simple spring with no more than simple addition, subtraction, and multiplication – you don't even need a calculator!

Fig. 17 *(opposite)*
Nomogram connecting spring rate; number of coils; wire dia. and D/d ratio. This chart is drawn for Regular Carbon Steel with modulus of torsional elasticity $G = 11.4 \times 10^6$ and includes the allowance K_2 of Fig. 4.

For any other material, modules = G^*, multiply the left hand "Rate" scale by $G^*/11,400,000$. (The figures in brackets are very approximate values for Phos. bronze wire, $G^* = 6,000,000$.) The chart has been prepared in Imperial units; S.I (metric) figures obtained by conversion.

Compression and Tension Spring Design 27

Special Note This is important. This nomogram, by its very nature, has to be drawn for *one particular value of the torsional modulus,* and Fig. 17 is worked out for the ordinary carbon steel spring wire, which has a "G" of 11.4×10^6. (Another way of writing 11.4 million – it means 11.4 with six noughts afterwards.) However, it can still be used for other materials simply by altering the scale of column A_1. For convenience, I give below the figures of Table I again.

Carbon Steel	G = 11 400 000lbf/sq.in.
Piano Wire	G = 12 000 000
Stainless Steel	G = 10 000 000
Phos. Bronze	G = 6 000 000

The difference between carbon steel and piano ("music") wire is small and can often be neglected, but for the others, it is prudent to make the correction.

There are two simple ways of doing this. Suppose we want a bronze spring having a rate of 9lbf/in. Method (1) is to say that an identical spring made from regular carbon steel wire would have a rate of 9 times the ratio of the two moduli, i.e. $9 \times 11.4/6 = 17.1$ and then use the nomogram for this figure. That is the best way if you need materials other than carbon steel rarely. For stainless steel, again at 9lbf/in., you would use the nomogram at $9 \times 11.4/10 = 10.3$. The *second* method is simply to pencil in a further scale alongside that for carbon steel – I have done this for a few values for bronze on Fig. 17. The new scale figures will be $G^*/11.4$ times those on the chart, where $G^*$is the modulus of the material being used. This is useful if you have a lot of springs in the second metal to design.

Note that this does *not* affect the load/stress nomogram, Fig. 16, as in that one we use the *actual* stresses, and these can be selected to suit. It only affects the "Rate". I will be going through the design of a bronze spring when we come to a few practice examples, but meanwhile let us summarize the fairly considerable amount of knowledge you have acquired so far!

Summary First, on the relationships between the load and various dimensions of a spring. Check each of these statements either against the formulas given at the beginning, or on Figs. 14, 15, 16 or 17, and make sure that you agree. In each case "vice versa" applies.
(1) For a given load and wire diameter – Increase in D/d ratio **increases** the stress.
(2) For a given load and D/d ratio – Increase in wire diameter **reduces** the stress.
(3) For a given stress and wire diameter – Increase in load requires a **smaller** mean coil diameter.
(4) For a given stress and load –
 (a) Increase in coil dia. requires a **thicker** wire.
 (b) Increase in wire dia. requires a **larger** coil dia.
 (c) Increase value of D/d requires **thicker** wire.
Second, on the rate and the various dimensions.
(5) For a given rate and wire diameter – increase in D/d requires **more** coils.
(6) For a given rate and D/d ratio – Increase in wire dia. requires **more** coils.
(7) For a given D/d ratio and wire diameter – increase in rate requires **more** coils.
(8) For a given wire diameter and number of coils – increase in D/d **reduces** the rate. Many of these

are fairly obvious, but it is helpful to have them on a piece of paper by you when trying to adjust a spring design.

Coil Spring Design Procedure

Before you start, look over the constraints you have imposed on the design – always try to keep these as few as possible. If you have but one size of wire available, need a specific load, at a certain length, a definite rate, and the coil diameter is limited by the space available, you are going to have very great difficulty in designing a spring, and it could turn out to be impossible!

Having checked the constraints, the steps are as follows:

(1) Make a first estimate of wire and coil diameter from Figs. 15 or 16, with, of course, the limitations of the mechanical design of the application in mind. Keep within $D/d = 5$ and 14 if possible, remembering that the larger the figure the "softer" the spring will be.

(2) Use the "Load" Nomogram – Fig. 16 – to refine this first estimate of D and d, remembering that you can, within reason, adjust the stresses. Keep in mind the need to avoid over-stress at full compression or extension; it helps to make sketches of the mechanical arrangement showing the different positions of the load.

(3) Use the "Rate" Nomogram – Fig. 17 – to determine the number of coils needed. Check (a) that there will be room for that number of coils at the loaded position, (b) that if it is a compression spring, the stress lies below the elastic limit (Figs. 5-8) when fully compressed or, in the case of a tension spring, that the hooks can be engaged without overstretching it. Make such adjustments to the design as may be needed.

(4) Go back to the Load Nomogram and check the stresses with the new dimensions.

It is normal to have to go through this procedure twice and in difficult cases several times, and it may even be necessary to change the material if the only dimensions which will fit the space available result in stresses which are too high. But, as I said earlier, there is *no* harm in using LOW stresses. You may be a bit put out at "not getting the right answer straight away" but, if you think for a moment, there are so many variables involved in spring design that it is really more a matter for surprise if it comes out right first time!

Important Note

Nomograms can be used with greater accuracy if size is increased. Regrettably they cannot be reproduced across two pages without loss of accurate alignment, and there are constraints which render the provision of fold-out pages impracticable. Tests have shown that twice-size copies can be made without significant distortion on modern dry photocopies with enlarging facilities, and this procedure is recommended to readers likely to make extensive use of the charts.

CHAPTER 4

Worked Examples

In reading this section, follow the steps yourself on the charts. This will give you physical practice and you will gain much more benefit from them. Use a very soft (B or BB) pencil sharpened to a flat chisel point to mark the reference line; the marks will then rub out easily. And as I mentioned before, use a transparent straight-edge. You must expect to get slightly different answers from mine, as all paper is liable to stretch a bit according to the weather, and it was − 7 deg. C outside when I did the work!

Example (1): *A bronze spring is required for the safety valve of a "toy" steam engine, to exert a force of 0.25lbf at an operating length of ⅝in. It must pass over an 8BA rod (0.087in. dia.).*
• Start with Fig. 14. This is drawn for carbon steel at 70% of the elastic limit, but we are to use bronze, and at a reduced stress due to temperature. A first look at Fig. 14 suggests that wire of 28 or 30 gauge might do. Look at Figs. 5 and 8 (which give the stress-diameter curves) and see that carbon steel can work at 93,000 and hot bronze at 37,000lbf/sq.in. The spring would exert a force of 0.25 × 93/37lbf if made of carbon steel instead of bronze, = **0.63**lbf.
• *Estimate the spring diameter.* Allow 0.006in. clearance on 8BA, then min. internal dia. is 0.093in. Guessed wire is 0.0148 dia. (28 s.w.g.) so that the mean dia. will be 0.093 + 0.015 = 0.108in.
• *Use this on Fig. 14,* and see that, in fact, 30's gauge wire is almost exact at D = **0.105**in. D/d is, by calculation, **8.5**. Check that this will pass over the 8BA rod. (Do this yourself!)
• *Turn to Fig. 16* but use the *correct* load of 0.25lbf this time. Align the figures across the two "B" scales and mark the reference line. Then from D = 30 s.w.g. on A_2 align across to read that the stress is **37,000**lbf/sq.in. Refer to Fig. 8 and see that this is acceptable.
• *Now ascertain the "Spring Rate".* Use Fig. 17, but remember that this chart is drawn for carbon steel. Guess a suitable number of coils. Your guess is as good as mine − I chose 20, which with the "dead" coil at each end makes 22 in all when calculating the compressed length later.
• Align from D/d = 8.5 to wire dia. = 30 s.w.g. on the "B" scales, mark the reference line. Then from 20 coils on A_2 read across to A_1 and find "rate" = 1.4lbf/in.

But this is for carbon steel, and we must allow for the different torsional modulus of bronze – see Table I (page 11). The true rate will be 1.4 + 6/11.4 =**0.74**lbf/in.
• From this we can calculate:
Compression to provide 0.25lbf is 0.25/0.74 = **0.34**in.
Free length, unloaded, is 0.625 + 0.34 = **0.965**in.
Fully compressed length is 22 × 0.0124 = **0.27**in.
Fully compressed load is (0.97 – 0.27) × 0.74 = **0.52**lbf.
• We must check that the spring is not overstressed in this condition of full compression. Go back to the load/stress nomogram, Fig. 16. You will see that the stress is about 77,000lbf/sq.in., which lies rather above the elastic limit for that size of wire on Fig. 8. *This must be reduced.*

The alternatives are:
(a) Reduce the length to about 0.9in. (i.e. wind the coils closer together). Then the fully compressed load will work out at 0.47lbf. (Try that for yourself.) Fig. 16 shows the max. stress to be 68,000lbf/sq.in. The length at the working load of 0.25lbf will occur at 0.9 – 0.34 = **0.56**in., or ⁹/₁₆in. This may be permissible.
(b) Increase the number of coils. Try this one for yourself, using, for example, 25 free coils (27 in all). You should find that the maximum fully compressed stress is about 65,000lbf/sq.in., and the free length about 1.03in.
(c) Arrange the mechanical design so that the spring can never be fully compressed. I try to do this in any case, but it won't prevent people from "trying the spring" and overstressing it between their fingers!

Of the above alternatives, I would recommend (b), as a low rate is desirable an a safety valve.

Example (2): A Tension spring is required to pass through 5/32in. bore tube and to exert forces of 1lbf when 4.75in. long, ½lbf when 3in. long, and a slight tension when 1½in. long. Find the maximum safe extension length if it is made of carbon steel.
• *Estimate the wire and coil dia.* from Fig. 14. Work on the maximum working load of 1lbf. (The chart is drawn to give the necessary factor of safety already.) See that 28 s.w.g. wire at 0.12in. mean diameter may serve, and that this gives an O.D. of 0.135, smaller than ⁵/₃₂in. Work out D/d = 8.0. From the load nomogram find the stress at load = 1lbf. This is about **91,000**lbf/sq.in. which, from Fig. 5, is safe.
• Turn to Fig. 17. From the given data, the change in force needed is 0.5lbf when the length changes by 1.75in, or 0.285lbf/in. If the load is 1lbf when the length is 4.75 we see that at the "almost slack" length of 1½in. the load will be 1 – [0.285 × (4.75 – 1½)] = **0.074**lbf – the "slight" tension asked for.
• *Estimate the number of coils.* Use Fig. 17. Align across the "B" columns at D/d = 8 and d = 0.0148in. (for example, 0.015). Mark the reference line and then project from "Rate" of 0.285, to find that we are off the scale, with perhaps 140 coils needed. But with this wire and a required minimum length of 1½in., there is room for only about **100** coils. (Work it out!) This is NOT a catastrophe, just an example of the sort of thing that may happen at the first trial design of a spring.

Proceed as follows:
• We know that we cannot have more than 100 coils, so try 93. Align from rate = 0.285 on A₁ to 93 coils on A₂ and mark the reference line. Then from B₂ at the wire dia. of 28 s.w.g., align across to B₁, finding a new D/d ratio of **9.2**.

- *Check for size.* D = 9.2 × 0.0148 = 0.136, and O.D. = 0.15in. This will still pass through the tube, so can be accepted.
- *Check for stress* on the load nomogram, Fig. 16, using the new D/d ratio, I make the stress to be about **102,000**lbf/sq.in. (Try it yourself.) From the stress chart for carbon steel, Fig. 5, you can see that the elastic limit is 132,500lbf/sq.in., so that at the working load the stress will be 102/132.5 = 0.77 of the limit; this is a good margin.
- *The final check* is the maximum extension. Here the stress will be AT the limit, so use 132,500 on the nomogram, Fig. 16, and you should find that the load is now **1.25**lbf. This means an extension of 1.25/0.285 = 4.39in. The minimum length is 93 × 0.0148 = 1.376in. (number of coils × diameter) so that the maximum extended length will be 4.39 + 1.376 = **5.76**in. (This excludes any hooks, of course.) So, the mechanism must be designed so that the spring can be installed without stretching it to more than 5¾in. (There is no way of limiting the "stretch" of a tension spring, which is inherent within the spring itself.)

This example was tightly constrained but, as you see, a satisfactory design emerged after just one additional trial, the spring being wound close coiled with **93** coils of **28** s.w.g. on a mean diameter of **0.136**in. (We shall be dealing with winding mandrel diameters later.)

Example (3): An axlebox spring is required to exert a force of 14lbf when ¾in. long and to pass over a 5BA (0.126in.) bolt. It is desirable that it be not more than 1in. long.

I propose to give the calculations in summary only for this one. Check the work yourself as we go along!
- *Fig. 15.* 18 s.w.g. selected; check mean dia. to clear 5BA, gives 0.18in. D/d is then 3.75. Rather tight; try D/d – 4, which D = **0.192**in.
- *Fig. 16.* Stress = 65,000lbf/sq.in. Safe. Max. desired length = 1in. sq. so that compression for 14lbf is ¼in., and rate = **56**lbf/in.
- *Fig. 17.* This rate needs about 17 coils, and there is not space for so many. Assume fully compressed coil length to be ⅝in., then total number of coils = 13, of which 11 will be active. Enter this number of coils – we are off the chart, but can extrapolate, as the vertical distance between "6" and "9" is the same as that between "60" and "90" on the scale. Rate comes out at **90**lbf/in. (Can be checked from the basic formula if need be.) Fully compressed length is **0.624**in.
- *From this rate* the compression to working load is 14/90 = 0.155in. and the "free length" will be 0.75 + 0.155 or **0.905**in. This is "not more than" 1in. long.
- *Load at coil/coil compression* (0.9–0.624) × 90 = **25**lbf.
- *Check Stress* when fully compressed, Fig. 16 = **90,000**lbf/sq. in.

This spring is a very tight one, both as to rate and D/d ratio, and would be difficult for the amateur to wind, though not impossible. It would be better made from piano wire, using the much higher stresses then permitted, and so reducing the "rate" and increasing the D/d ratio.

Example 4: Four valve springs of a very old paraffin engine from the scrapyard have mean diameter 1.18in., wire about 12 s.w.g. (0.104in.) wound with 10 free and two dead coils. 2½in. free length. Are they safe at the maximum load and what is the rate?

(This type of calculation is not uncommon – most junk-boxes contain scrap springs!)
• Work out that D/d is 11.4, and the fully compressed length is 1.25in., very nearly.
• *Fig. 17.* Enter the figures as before, to find Rate = **9.5**lbf/in. Hence max load at full compression is:
(2.5 – 1.25) × 9.5 = **11.25**lbf.
• *Fig. 16.* Enter the figures, to find that the stress is **25,000**lbf/sq.in.

The fact that this cheap engine was designed c.67 years ago justifies the low stress for a valve-spring!

To complete this section, try a bit of "homework"! Find a spring from your box which is of such a size that you can measure the load and rate. Use the charts to CALCULATE what these should be and compare with the measured figures. (N.B. Don't use a modern engine valve-spring, for these are usually made of chrome-vanadium or other alloy steel. Use one of "model" type.) Then design one to do the same job but (a) of larger O.D. and (b) made of bronze. In short, get in as much practice as you can NOW. Then, when later on you actually need a spring, the procedure is more likely to come back to you.

Reproducing the Charts

If you are likely to design springs fairly often you may not wish to mark the pages. You can photocopy the charts on a *plain paper copier* with reasonable accuracy so long as you keep the pages dead flat while doing so, but Thermofax or other heat-type two-sheet copiers may not be so accurate. Figs. 5 to 8, and Figs. 14 and 15 don't matter in this respect – they will reproduce "near enough" – but the two nomograms should be done as carefully as possible. You should find that a local "Reprographic" firm will copy them for a few pence, and I suggest that you have them done on stiff white card rather than paper. If you can draw up logarithmic scales, you COULD copy them manually, but the positioning of each scale and the spacing of the verticals is *very* important. Unfortunately, the preparation of 4-variable nomograms from scratch does involve a bit of math, as well as making a different log scale for each variable. Photocopying will give adequate results.

Accuracy

There is a tolerance on commercial spring wire of about ± 1 % in larger sizes, which may rise to 3% (or 0.0005in.) for wire below about 24 s.w.g. (for example, 0.02in.). The rate at a given load depends on the *fourth power* of the wire diameter, so that variations in rate could be from 4 to 15%. This can be overcome by *measuring* the wire and using this dimension in the nomogram. Then, provided a reasonable working stress is assumed this tolerance need not be of importance. The IMPERIAL figure on the nomograms will be within 1%, metric derived to within perhaps 1% on top of that, but "wire gauge" markings are not so close. Errors in *reading* it depend on how careful you are, but the stress read off should be within 5%.

Coil diameter errors can be rather more, as we shall see later when we consider spring-back during winding. A 1% tolerance of wire diameter makes about 4% difference in Scale (or Rate) and an error of 1 % in coil diameter will cause a Rate error of about 3%. Further, it is not easy to wind an exact number of active coils. These tolerances, together with any slight errors in using the nomogram, *may* cause the actual Rate to differ from that calculated by as much as 10%. Fortunately, it is rare that model engineers need sets of springs accurately matched in rate, but where this IS needed, then all you can do is, first, measure the actual wire diameter; then, wind a trial spring and measure the true mean diameter; next, repeat the nomogram exercises to adjust the number of coils accordingly, and, finally, take great care when winding to get the exact number of coils needed.

In general, however, springs designed using these charts will perform much better than those arrived at by trial and error or, worse, by "stretching and squeezing" lengths of stock spring coil!

CHAPTER 5

Winding Coil Springs

Spring Back
When we wind a coil spring, we usually find that although it lies close to the mandrel during winding, the coils slacken off a little as soon as we release the winding tension. This is an inevitable consequence of the winding process. The act of "winding" is, in essence, a bending action which takes the majority of the fibers of the wire beyond the elastic limit, so that it takes a permanent set. Look at Fig. 19. The diagram at (i) shows a bar bent over a mandrel "S", the bending action represented by the force "W". The stress within the bar will be a maximum at XX, where the bar loses contact with the mandrel. This stress is shown at (ii). At the top face, the stress is negative – a tensile stress – and this diminishes in magnitude across the bar, being zero at the center or neutral axis. It then becomes positive – compression – increasing till we reach the bottom face of the bar. In the diagram, the compressive region is shown shaded.

Increase the bend further, as at (iii); the outer fibers now exceed the elastic limit, but there is still a region, between "B" and "C", where the stresses are still *below* the elastic limit – the material will be "elastic" in this area – though plastic flow has occurred between "A" and "B", and between "C" and "D". See (iv). This means that the fibers AA to BB in (iii) will have stretched permanently, and those between CC and DD will have shortened, again permanently. Now remove the bending forces, as at (v). The locked-up stresses in the material on either side of the neutral axis 0-0 In diagram (iv) are unbalanced, and will try to re-establish a condition of equilibrium within the elastic region – from "B" to "C". This condition is shown at diagram (vi), and you will see that the effect is to lay a permanent *compressive* stress in the upper fibers of the bar, and a *tensile* stress in the bottom ones. Now, there can be no stress without a corresponding STRAIN, or deflection, and the result is that the bar springs back from its original curvature as shown by the dotted lines in diagram (v).

The amount of this spring-back will depend on the elastic limit of the material and on the position of the boundaries BB and CC. On the face of it, we ought to be able to calculate what this effect will be, but unfortunately this is fraught with difficulty. My diagrams are a considerable over-simplification of

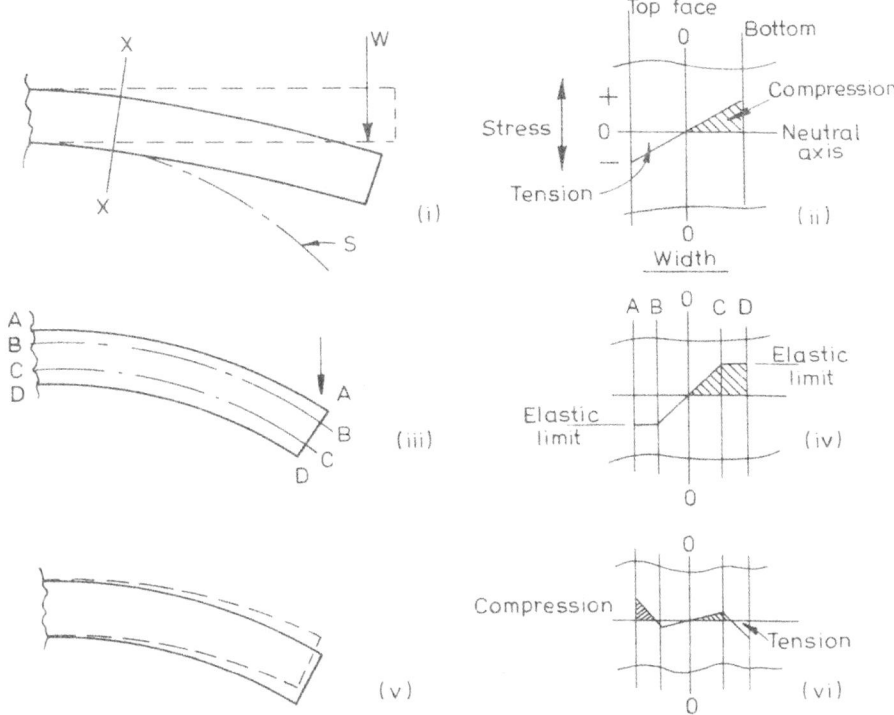

Fig. 19 *How locked up stresses cause "spring back" after winding. (i) Bar bent within the elastic limit. (ii) Shows the stress pattern. (iii) and (iv) Stress pattern when the bar is bent beyond the elastic limit; the materials are still "elastic" between "B" and "C". (v) and (vi) Stress pattern causing the spring-back after the load is removed. Note: Compressive stresses are shown shaded.*

the phenomenon. To begin with, the elastic limit in tension is less than that in compression; the transition is not sudden as I have shown it and even within the plastic areas the metal can still carry a small stress. Most of all, a spring wire does not behave exactly like a beam, for while a beam remains relatively flat when fully loaded we have put an enormous bend into the wire, and the "simple theory of bending" applicable to beams no longer applies. Overall, the very act of bending a *spring material* beyond the elastic limit work hardens it a little more. So, there is no simple way of forecasting the effect other than by experiment. We have to wind a trial coil and see what happens.

However, there is available just a little experimental data which helps – at least in giving us a guess at the size of mandrel on which to wind the trial spring! Fig. 20 shows the average of the results of a large number of trials made on PIANO WIRE. The vertical scale shows the ratio – Increase in coil dia. divided by the wire dia. These showed that the D/d ratio has an effect for wires above about 0.1mm dia. and that the spring back tends to increase with an increase in D/d ratio – as we should expect, for the larger the radius of curvature at "S" in Fig. 19, the more metal is left within the elastic region between "B" and "C". The other factor which can have an effect is the tension put into the wire between the bobbin and the mandrel, but with hand winding methods in the lathe, this is unlikely to be serious. Finally, the effect will be more pronounced as the strength of the wire is increased. Fig. 20 is for piano wire, and we should expect ordinary spring wire to be less affected (justifying the rule used by some that increase in coil dia. due to spring back is equal to the wire diameter). Spring back will be even less when using phosphor bronze.

Trial Coils

Find a piece of round stock which is as near as to the required mandrel

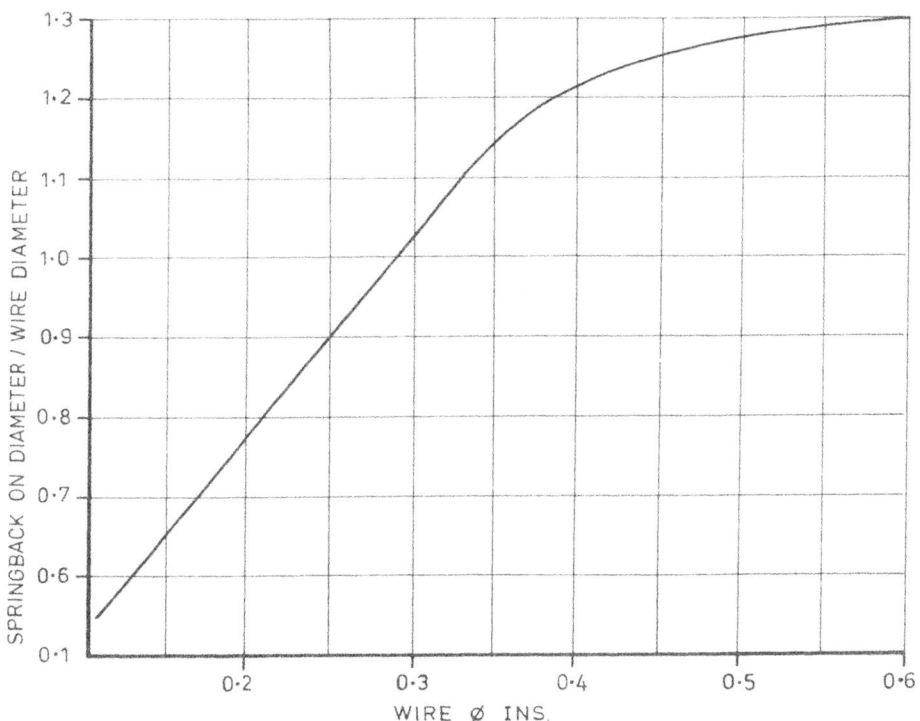

Fig. 20 *Graph of spring back against wire diameter for PIANO WIRE. The vertical scale shows the ratio between the increase in coil diameter and the wire diameter. For ordinary carbon steel wire use 80% of these figures, and for phosphor bronze, 60%.*

diameter as estimated from Fig. 20. If using regular carbon steel wire, use about 80% of the figure shown, and 60% for phosphor bronze. There is no need to be precise; just choose a diameter "as near as possible" without making one. Wind, by hand, 1½ to 2 coils putting in as much tension as you can manually. Then measure the O.D. of the coil after releasing it from the mandrel. You know the diameter of both wire and mandrel, and can thus work out the diameter to which the final winding mandrel must be machined.

The alternative approach, of course, is to use standard diameter stock as you have for both test and winding mandrel, and after checking the actual mean coil diameter after spring back, refer to the nomograms to ascertain whether or not the spring will provide a "near enough" spring force and rate. We have to preserve some sense of proportion over this matter. Most springs whose in situ load is important have means of adjustment provided; with others, it doesn't matter a great deal if the O.D. is slightly larger than designed (though it may be important that a spring doesn't bind on, e.g. on a valve-stem, though here the spring back will tend to ease it rather than the reverse). I find that less than one in a dozen of the springs I actually *wind* need the meticulous attention described above; indeed, for many of my springs I find that I can "adjust" lengths cut off from commercial "stock lengths". (I shall be dealing with such adjustments later.)

Hand Winding

It is possible to wind springs having small diameter coils of thin wire just by hand, though it helps if the mandrel has an enlarged end to ease holding, and a hole to capture the end of the wire. It is vital that the wire be absolutely free from even mild kinks as it is impossible to apply sufficient tension to eliminate these when winding – besides which there is the risk of physical injury as well. By using a combination of repeated twists of the mandrel in one hand and a sort of "wrapping" motion under the thumb of the hand holding the wire a tolerable spring can be wound, especially with phosphor bronze or brass wire. There is little problem in spacing when winding close-coiled tension springs; for compression springs, which are fairly close-coiled, the trick I used when making radio coils can be applied – interwind a second coil of soft copper wire between the spring coils, of a gauge to give the desired spacing. Hand winding is a useful technique for the odd spring that doesn't matter very much.

Machine Winding

The machine, in our case, is usually the lathe, with one end of the mandrel held in the 3-jaw, the other on the tailstock center. Some writers suggest capturing the "start" end of the wire under one jaw of the chuck, but this is not a good idea; apart from throwing the mandrel out of axis it brings the start of the wire too near to the bulky chuck. (There is, in fact, much to be said for winding from the tailstock towards the chuck.) Wherever possible, I drill a hole in the mandrel or, with larger coil diameters, clamp the wire end with a jubilee clip or similar. With close-coiled springs, it IS possible to traverse by hand, though not really to be recommended. If this method *is* used, then I suggest that you (a) wear leather gloves and (b) run the wire round a small bobbin under your thumb, to get a more even tension.

Left, **Fig. 21** The "ACRU" spring-winder. A, Handle; B, calibrated wedge-washer; C, Tension washer; D, Tension adjuster; E, pivot bolt; F, locknut for adjusting B; G, washer. **Above, Fig. 22** The ACRU in action. It is preferable to have the hand tee-rest to support the tool. For close-coiled springs, the ACRU is used with the wedge-washer on the other side of the fed wire.

Though a bit more trouble it is far better either to make up a spring-winding device, or to use the change-wheels and guide the wire from the saddle.

A Commercial Winder

A commercial *spring winder* which I obtained almost 40 years ago is the ACRU, shown in Fig. 21. This comprises a hollow handle, about 1in. diameter, through which the wire is passed, thus reducing all risk of injury from cut ends. The wire passes between a fairly hard fiber washer "C" and a disc "B", and the grip of the washer can be adjusted to alter the winding tension. The disc "B" is graduated in thickness (it is in the form of a circular wedge) from about 0.05in. to 0.175in. This is adjusted to a position which gives the required coil spacing, one side of the wedge bearing on the coil already wound, and the other, of course, on the feeding spring wire. In addition, the O.D. of the wedge-washer bears on the mandrel and helps to relieve it of some of the bending forces. The device can be used either way up, so that both left- and right-hand helices can be wound. Fig. 22 shows it in operation. The coil is a bit ragged, because it "lost its grip" while I was setting off the camera! The device does need a little practice, and is best used with the hand rest underneath to give some support, but provided that the initial coil is wellformed (best done by pulling the lathe round by hand), you can wind springs on middle back-gear effectively. It is made by the ACRU ELECTRIC TOOL Co., Ltd. Universal and modern spring winders are available through a wide range of companies. It can, of course, be used for winding tension springs as well, simply by turning it over so that the wedge-disc runs ahead of the coils.

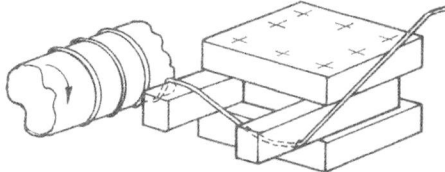

Fig. 23 *Arrangement of two wood or brass guide-bars in a 4-tool turret for winding springs using the lathe leadscrew for pitch control.*

Winding with Power Traverse

This is by far the best way to achieve correct spacing of the coils. The change-wheels must be set to give the required pitch of the coils as when screwcutting (or, for fine pitches, set as for normal traversing) and the tumbler reverse used to feed either from head- or tailstock as desired. In this connection, it will be realized that feeding from the headstock gives a left-handed helix, and vice versa. Some form of guide is, of course, needed and Fig. 23 shows the sort of thing I use in the 4-tool turret. The "bar" nearest to the mandrel is a piece of hardwood (boxwood in this case) with a very small notch over which the wire passes. It can

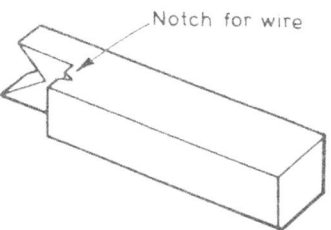

Fig. 24 *This type of guide is preferred when winding on smaller mandrels. The vee at the front must be set at correct center height.*

be of brass if the wire is exceptionally stiff. There is a second bar of wood set in the opposite side of the turret, and the wire is passed under this and then upwards. The combination of the two bends passing over the bars increases any tension applied by the hand (gloved), and the magnification can be increased by leading the wire onto the outer bar at increasing vertical angles.

Fig. 24 is a device which allows heavier wire to be wound onto smaller mandrels. The vee-notch embraces the mandrel and gives it some support. Again, there is a small guide-notch to align the wire. This type of guide is best made in metal if the wire is above about 20's gauge, brass for preference. It must be set on packing to get the mandrel vee exactly at center height, of course. In both types, the application of sufficient tension can sometimes be a small problem, but one which can easily be overcome by using the ACRU spring winder as well. This can be adjusted to provide as much or as little tension as is necessary, and if the wedge-disc is held against the front bar of Fig. 23 (the rear bar can be dispensed with) or the end of the guide shown in Fig. 24, the job becomes easy.

In winding a compression spring (Fig. 25), the procedure is as follows. The mandrel is prepared and the, change-wheels set to give the desired pitch. The guide is set up in the toolpost and the saddle brought to a point near the wire anchor – a clip in the photo. The leadscrew is rotated until the half-nut can be engaged. With this released (and the tumbler reverse in the correct position), the chuck is rotated *backwards* for about 3 turns and the wire then engaged with the clamp – or passed through the hole if that is used. It should be bent round to be sure of a secure hold. I should have said that the middle

Fig. 25 *The arrangement of Fig. 23 in use winding a compression spring. Note the first and last few coils close-wound to ease flattening the spring ends later.*

or slow backgear will be used. When all is in order, the lathe is started and will pull in the wire. The first few coils will be close-wound, but as soon as the leadscrew indicator comes round to its mark, the half-nuts are engaged and from then on the coils will be wound at the correct pitch. When sufficient number have been wound, the half-nuts are disengaged and the lathe allowed to wind another two or three close coils. The wire can then be cut and the spring removed. The object of the few close coils at each end is to ease the subsequent end-preparation. Most compression springs call for the ends to be "ground flat" and with open winding this is not easy. We shall deal with this later.

Tension springs are wound in exactly the same way – Fig. 26. The main problem here is finding a change-wheel train which will match the wire diameter. Fortunately, this is not critical. The wire can run on at a slight angle and still not override the previous coil. (It is desirable to set the pitch slightly LESS than that designed rather than the other way.) However, if winding a long spring the progressive error *can* become

Fig. 26 *Guide as in Fig. 24 in use winding a tension spring. In this case, the wire end has been captured through a hole in the mandrel.*

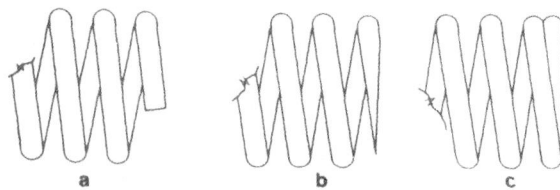

Fig. 27

a nuisance, and in such cases, it is necessary to make slight adjustments to the topslide to keep the wire properly aligned. As in the case of compression springs, a few extra coils must be wound at each end to allow for making the end loops.

Initial Tension

Most tension springs are wound with "Initial Tension". That is, it requires a definite force to be applied before the spring begins to stretch at all. Contrary to common belief this feature is *not* achieved by winding the spring with the wire at an angle to the circumferential line of the previous coil. True, this does impart a small tendency for the coils to nestle close to each other, but it does not cause any measurable initial tension. If you think back to the beginning, you will remember that the spring derives its ability to exert a force from the *torsional stress* in the wire. To obtain an *initial* tension, we must wind the spring with a locked-up torsional stress in the wire. This is done by *twisting* the wire as it is laid onto the mandrel, the twist being in the direction which will tend to cause the coils to press one against the next.

This is easily done on proper spring-winding machines, and both calculation and experience can be used to determine the amount of twist per revolution of the mandrel needed to provide the designed initial tension. For the model engineer, dependent on a lathe, this is not unfortunately practicable. However, it is possible to wind a spring with a small, though unquantifiable, initial tension if one of the ACRU type spring winders is available, or one similar. Referring again to Fig. 26, if the wire is fed first through the ACRU and then over the guide system, with the nose of the winder against the end of the guide; and if the winder is rotated slowly anticlockwise as the wire is wound on, then there will be SOME initial twist put into the wire.* How much cannot be estimated, as although the winder is turned about one revolution for each rev. of the mandrel, this does not guarantee that the wire itself will be twisted by the same amount. However, it is worth a bit of experiment, even if only for fun. The one point on which care is essential is that the speed at which the winder is rotated MUST be steady and uniform. If not, then the degree of initial tension will vary along the spring, and in service one part will start to stretch before the rest, and may well be overstressed at full extension.

As with all other manual operations, the effective winding of springs does require some practice – I find that I have to have one or more dummy runs if I have not wound a spring for some time.

---

*(The twist should be *clockwise* if the spring is being wound from the tailstock end towards the headstock, with the normal rotation of the lathe.)

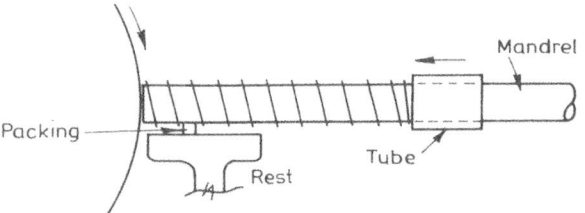

Fig. 28

Practice with soft iron or copper wire, which you can unwind, straighten, and use again. Then try the odd spring with phosphor bronze wire before going on the stiffer carbon steel type. Piano wire is the most difficult, as it has a very high tensile (i.e. in bending) strength. Although their use undoubtedly interferes with manipulative dexterity, I do recommend the wearing of strong leather gloves – with piano wire especially. And, to avoid the devastating effects of flying wire ends *ALWAYS ENSURE THAT YOU HAVE MORE THAN ENOUGH WIRE TO COMPLETE THE SPRING.* If the end of the wire catches up with you during winding, it can give a very nasty gash, and in the place where it does most damage, on the inside of the wrist.

Spring Terminations
Compression springs can, of course, simply be cut off as shown in Fig. 27A. This is not very satisfactory. First, the whole load comes on the very end of the wire and this may cause overstress. Second, and perhaps more important, the point load on the end of the wire causes the whole spring to cock over. A better arrangement is to grind the end of the spring as shown at 27B. This spreads the load over about half the coil; a much better arrangement. The ideal is the "closed end coil ground flat" shown in Fig. 27C. You will remember that I recommend the winding of a couple of coils at close pitch, and this is one of the reasons. (There is another type of termination, in which the final coil is "bent flat", but this is more typical of large hot-wound springs than of the type made by model engineers from hard-drawn wire.)

The flattened ends seen in "B" and "C" are, of course, made by grinding the end. In older books, you may find the advice that if this grinding is fierce enough the end coil will rise to red heat, and the pressure against the grindstone will cause the end coil to flatten, producing type "C" termination. This *can* be done, but it is a bit hit-and-miss, and there is a risk of grinding off more than was intended. The winding method which I have suggested earlier is a far better way. However, in either case the end must be ground, and this can cause problems, especially on the weaker type of spring. The trick is to set the spring on a dummy mandrel, fairly well fitting, with a piece of tube slipped on behind the spring. The whole is set on the tool-rest of the grinder with the mandrel almost touching the wheel and then light pressure is applied using the tube. Rotate the spring as you grind and this will help to ensure an even surface. *Don't* let the end of the spring get red-hot, and don't apply too much pressure. Naturally, it helps if you can offer the spring to the flat face of the wheel, but if you are careful, grinding from the circumference will give a reasonably flat end. *(See Fig. 28.)*

Tension springs. Fig. 29 shows a few examples of the terminations for extension (or tension) springs. Those at A, B, and C have open-ended loops for slipping over the anchor-point. The side view of A shows how the wire is preferably bent, with a reasonable radius to the bend, and also indicates how the centerline of the loop can be brought over the axis of the spring. B is used when the anchor is a peg of diameter comparable to that of the spring, while C is intended for smaller pegs. At D, we have two examples of closed loops, large and small, and at E, an arrangement where the loop is remote from the end of the spring. In forming this type, do *not* make the extension from wire unwound from the spring if this can by any means be avoided. Leave sufficient wire unwound in the first place, and try to ensure that you wind the correct number of coils, so that there is no need to unwind any. F is the type of termination which I prefer, consisting of two coils bent over. The eyes are, of course, all subject to bending stresses and can easily be overstretched during assembly. F will reduce the risk of breakage considerably.

Rate Adjustment

There is little that can be done to adjust the rate of a compression spring and not much that can be done to adjust the load at a given extension. No harm will come from a slight "scragging" of the spring – that is, overloading the spring slightly in compression so that it takes a permanent set, and shortens as a result. The mathematics are too involved to go into here, but this process can, if carried out within reason, actually improve the

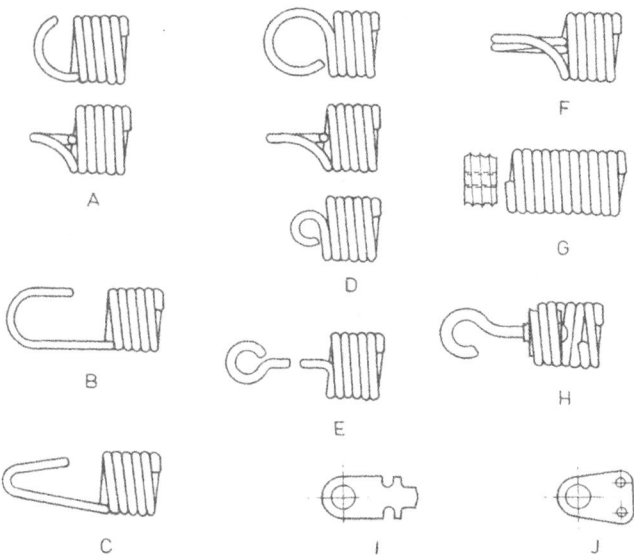

Fig. 29

performance. However, any attempt to *stretch* the spring to lengthen it must be **avoided at all costs.** The effect of such a procedure is to "lock up" stresses within the wire, which *add* to those imposed by the normal spring action, and this can, and often does, result in premature failure of the spring. With compression springs, it pays to get things right in the first place, and if either rate or load capacity falls outside the acceptable tolerance of the design load, then it is best to wind a replacement.

With *tension springs*, we have more scope. If the spring is wound in the first place with an excess of coils, we can correct by removing coils to increase the rate, or leave them on to reduce it. However, there are ways of designing adjustment into the end terminations, as shown in Fig. 29 G to K. Both G and H feature a threaded bobbin, the pitch of the thread on the O.D. being very slightly larger than the pitch of the coil. G has a threaded hole through, to which any anchor can be attached, while H has a hook brazed or riveted in place. In both cases, the effective length of the spring, and hence the number of coils, can be altered by screwing the bobbin in or out of the coil. The external thread (which should have a radiused profile corresponding to the wire) being slightly longer in pitch achieves a reasonable grip of the spring, but if absolute security is required than a clip can be put on to grip the coils. This type of end-fixing is the sort I should use for things like governor springs and the like. Incidentally, type G *could* be adapted for use on a *compression* spring having a Fig. 27A termination.

I and J are useful terminations in their own right, apart from the adjustment facility they provide; anyone having a suitable hole-punch can knock up type J out of thin sheet metal in a few seconds, much quicker than trying to form end loops. The "as wound" end of the spring is first trimmed off with a clean cut, and the first coil then wound into the two notches at I or the holes at J. Adjustment is made by winding more of the spring through. These ends are suitable only for small, light springs – about up to 5/16in. dia. × 20's gauge – though I have a commercial spring of type I which pulls a load of 12lbf, ¼in. coil dia. and about 19's gauge wire. There is, of course, no reason why the termination should not be made with a threaded rod brazed onto the spring fixing. The latter should be of good quality steel and of a thickness not less than the wire diameter. Note that the tongue seen on I is essential, and it should fit fairly snugly inside the coil.

Coil Springs of Rectangular Wire

Model engineers are unlikely to use other than round wire for their springs, but it may be of interest to have a few words about square or rectangular wire springs, as they are not uncommon on some prototypes. The effect of using (e.g.) square wire instead of round of the same diameter/side is first, to permit a slight (perhaps 5%) increase in load for the same stress, and, second, to make the spring much stiffer – the RATE can be increased by 25 to 30%. There are many and various analyses available for rectangular wire and no doubt in this computer age there are precise "finite element" analyses which will give an exact result. However, such precision is not necessary for models, even if the model engineer has access to a mainframe computer, and I find, on the very rare occasions that I have needed such springs, that the following formulas give reasonable results.

For LOAD, $W = \left\{ \dfrac{2.b^2.t^2 f_s}{D(3b + 1.8t)} \right\} \times k'$

For RATE, $R = \left\{ \dfrac{Gb^3 t^3}{2.62 D^3 n (b^2 + t^2)} \right\} \times k''$

D = mean coil dia.
b = radial breadth of wire
t = axial thickness of wire
f_s = shear stress
W = load
k' = see below
k'' = see below
n = number of coils
R = Rate

k' and k'' correspond to the corrections needed for round wire shown in the graphs at Fig. 4, page 9, but in this case, the base scale will be D/b, instead of D/d. For *square* wire, of course, b = t and the formulas become much simpler.

The easiest way to design a spring with SQUARE wire is to use the charts as for round wire, but with a load 5 to 6% *less* than that required, and a rate perhaps 27½% *greater*, and then check and refine the design using the formulas above. There is no easy way to design one with rectangular wire, I'm afraid.

The main benefit of using square or rectangular wire is that we can get a much *stiffer* spring into the same space.

CHAPTER 6

Leaf Spring Principles

Springs in Bending – Flat and Laminated Springs
Again we are going to be faced with "formulas," and rather a lot in the next few pages! This is because we are dealing with several types of spring, each of which behaves somewhat differently. Don't be dismayed! First, I shall be bringing most of them down into chart form. Second, some of them are needed only so that you can follow the argument – you won't have to work them out. Third, they look a lot worse than they really are; they are, after all, only MODELS, mathematical models, of the real thing, and if you can build a working model of a "Black Five", you should have no problem with these formulas. However, if such things do cause you great difficulty, try reading them in *words* instead of the symbols – thus, instead of f = W/A, read "Stress equals load over area", and it will make more sense! The charts will make calculation very easy, but they will be of even more benefit if you can understand WHY they are the shape they are. So, persevere and try your best with them.

A **flat spring** is no more than a special form of BEAM, but while a beam is usually designed to have minimum sag a spring *must* sag – "deflect" – by an appreciable amount. (Though most are designed to be almost flat in the working load position.) The simplest form is the *cantilever,* Fig. 30. The bending effect of the load – known as the **bending moment** – is clearly the lever effect and at any section, for example, a-a, will be W × L. It is immediately obvious that this effect or moment increases as we get towards the fixed end, and the maximum value will be W × L. The lower diagram in Fig. 30 shows this graphically, and the triangle is known as the **bending moment diagram** for the beam or spring. We shall see in a moment that the stress at any point depends on the bending moment, so that it is clear that the stress will be a maximum at the fixed

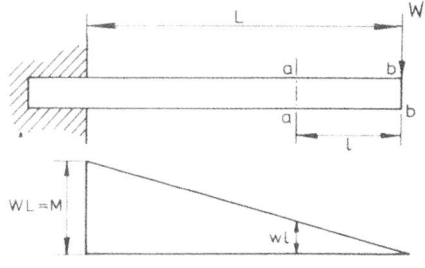

Fig. 30

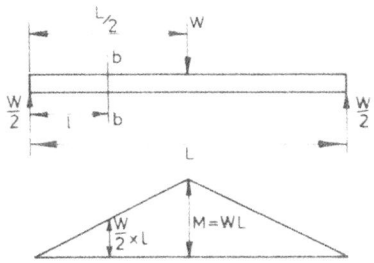

Fig. 31

end. However, there will also be a *shearing effect* at b-b, and the beam section must be sufficient to cope with this even though the bending stress here is zero.

The more usual form of spring is that shown in Fig. 31, supported at A and B, loaded at W in the center. It can be shown that the bending effect is exactly the same as that on two cantilevers with the (notionally) rigid fixing at W. The Bending Moment here is L/2 × W/2 = WL/4, and the diagram is again a triangle, but as shown in the lower part of Fig. 31. At any point on the beam, for example, b-b, the moment will be ½W × l, where l is the distance from the support; M_{aa} = Wl/2. The unsymmetrical beam is shown in Fig. 32.

Bending Stresses

If we imagine a beam made up of a number of thin planks, securely nailed

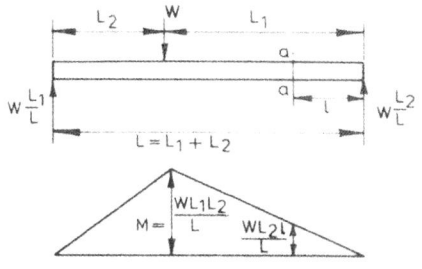

Fig. 32

together at each end, and then subjected to bending (Fig. 33), the whole issue will take up a curved shape – we assume that this is an arc of a circle, for what it is worth. It is fairly clear that the top plank will be shortened, and the material will be compressed, and that the bottom one will be stretched – in tension. The center plank will not (if they are thin) change length at all, and will be unstressed, and those in between will suffer varying degrees of tension or compression depending how far they are away from the center one. The stresses in the various planks can be shown graphically, as on the right-hand side of the diagram. As shown it represents an "infinite" number of planks, and the plus sign indicates tension, the minus sign compression. Each layer of the beam carries a different stress, with the outermost ones bearing the highest.

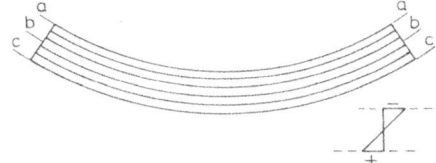

Fig. 33

Now, we are only concerned with that maximum stress, for if the spring material will withstand that, all the rest will be happy. But we can't just divide load over area, for the actual *load* in each layer is different. It depends both on the width of the load diagram on the right of Fig. 33 and also on how far the layer is from the center of the beam or spring. We have to use a concept (a "Mathematical Model", that's all) called the **section modulus**, which takes these factors into account, denoted by the letter "Z". The *maximum* stress is then given by:

$$f = \frac{M}{Z} \qquad (7)$$

The section modulus obviously depends on the shape of the cross-section of the beam, but for the usual rectangular section is given by:

$$Z = \frac{BD^2}{6} \qquad (8)$$

Where B is the Breadth of the section
D is the Depth (thickness)
This is, fortunately, not difficult to work out even if you have to do it by hand!

Units
You must, of course, use consistent units. Older readers will, like me, measure W in pounds, L in inches, B and D in inches, and so get M in lbf/in. and the stress in lbf/sq.in. (The "unit" for

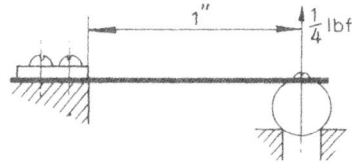

Fig. 34

"Z" is inches³.) If you work in ISO Metric Units, then W must be in Newtons (it is a *force*) L, B, and D must be in millimeters and the stress will come out in Newton/mm². (Obviously, for a large bridge beam you might use tons or Tonnes, meters, etc., but not for model springs!) Let us see how it works out.

Example
Fig. 34 shows an unorthodox but not impracticable safety-valve. The steam load on the ball is ¼lbf, and you have some 28 s.w.g. steel spring strip, ⅛in. wide. Will the stress be reasonable? (28 s.w.g. is 0.015in.).

Bending Moment,

$$M = W \times L$$
$$= 0.25 \times 1$$
$$= 0.25 \text{lbf.in.}$$

$$\text{Modulus, } Z = \frac{BD^2}{6}$$

$$= \frac{0.125 \times 0.015^2}{6}$$

$$= 0.0000047 \text{in. units}$$

Stress, $f = M/Z$

$$= \frac{0.25}{0.0000047}$$

$$= 53,190 \text{lbf/sq.in.}$$

Now check the shear stress

Shear area, $B \times D = 0.125 \times 0.015$
$$= 0.00188 \text{ sq.in.}$$
$$f_s = W/A = 0.25/0.00188$$
$$= 13 \text{lbf/sq.in.}$$

A safe maximum working stress in bending would be 75,000lbf/sq.in., and perhaps 35,000 in shear for tempered carbon spring steel, so that the dimensions are satisfactory in this respect. The shear stress can be ignored.

Deflection
As I have already said, the loaded beam or spring deflects into an arc of a circle. Clearly this will depend on the stress, and on Young's Modulus "E" (the ratio of Stress/Strain, varying for each material). We have to work out how much each fiber will stretch or compress, and this will depend again on how far it is from the center

of the beam. (In the textbooks, this center is called the "Neutral Axis", by the way.) It is possible to do this, but it is rather a long-winded operation, and in any case, it is usually more convenient to be able to work out the deflection directly from the LOAD, instead of having to work out the stress in each fiber, first. To do this, we make use of yet another "mathematical model" which equates to the effects of *all* the fibers in the beam or spring. This is called the **moment of inertia** of the section. (Though the odd "academic" of readers may object that it ought to be called "Second Moment of Area"; fair enough – it is that, too!) This is denoted by the letter "I", and leads to the very simple triple relationship that:

$$\frac{f}{\tfrac{1}{2}D} = \frac{M}{I} = \frac{E}{R} \qquad (8A)$$

Where R is the change in the radius of curvature. D is the depth of the beam. I, like Z, depends on the shape of the beam, but for the usual rectangular section is given by:

$$I = \frac{BD^3}{12} \qquad (8B)$$

(For other sections, both I and Z can be found by reference to Engineer's Pocketbooks or textbooks on Engineering Design. I = Z × D/2.)

Now, we are not in the slightest bit interested in the radius of curvatures as such, but rather in its effect on the straightness of the beam – i.e. in the maximum vertical *movement*. This occurs (in all except some unusual cases) directly under the load, and I give the formulas for these standard cases below. It is important to remember, however, that the deflection is *really a change in radius of curvature so far as the beam is concerned*. In the case of the leaf spring, for example – really a compound beam – the vertical deflection of the very short bottom leaf is very small relatively to its ends, but the change in its radius of curvature is exactly the same as the change in radius of curvature of the long top leaf.

I won't go into the derivations, but here are the expressions for deflections of the three types of beam or spring: For Fig. 30, Cantilever.

$$\S = \frac{WL^3}{3EI} \qquad (9)$$

For Fig. 31, Simply supported.

$$\S = \frac{WL^3}{48EI} \qquad (10)$$

For Fig. 32, Load offset. Hold your hats – fortunately this is a rare one!

$$\S = \left[\frac{WL_1 L_2 (2L - L_2)}{27EIL}\right] \sqrt{3L_2 (2L - L_2)} \qquad (11)$$

I include the last just for completeness – one never knows when some prototype will turn up with asymmetrical springing like that! Anyone willing to tackle such a model will undoubtedly be equally willing to work it out!

Before trying an example, note that the typical values of both working stresses and Young's Modulus of Elasticity for the more common materials are given in the table opposite.

Example
To ascertain the deflection at the end of the spring in Fig. 34.

Table III

Material	Maximum Working Stress lbf/sq.in.	Young's Modulus, E Inch Units
Carbon Spring Steel, Tempered	75,000	29,000,000
70/30 Brass, CZ106, Spring Temper	28,000	16,000,000
5% Phos. Bronze, PB102, Hard Temper	70,000	15,400,000
Beryllium Copper, CB101, WH Temper	80,000	18,000,000
"TUFNOL" (Fine Fabric Laminate)	8,000	1,150,000 approx.

Note the qualities of beryllium copper, which will work at the same or slightly higher stress than steel but will provide some 60% more deflection – helpful when "scale effects" are met with.

$I = BD^3/12$
 $= 0.125 \times 0.015^3/12$
 $= 0.125 \times 0.000\ 0034/12$
 $= 0.000\ 000\ 0354$

 $= WL^3/3EI$
 $= 0.25 \times 1^3/(3 \times 29\ 000\ 000 \times 0.000\ 000\ 0354)$
 $= 0.25/(3 \times 1.0266)$
 $= 0.081$ in.

(Note that the normal pocket calculator will not handle the long numbers involved, so either do it manually or use $E = 29 \times 10^6$ and $I = 0\ 0354 \times 10^{-6}$, if your calculator will handle powers. The 10^6 and 10^{-6} can, of course, be cancelled out.)

If this was felt to be too large a deflection it could be reduced either by shortening the length of the spring or by using thicker material. In both cases, the "cube law" would operate and in both cases the stress would be reduced. Reducing the length to $7/8$in., for example, would reduce deflection to about 0.051 in. and the stress would go down to about 46,500lbf/sq.in. Try the effect of changing to 26's gauge (0.018in.) for yourself.

"Springiness"

The formulas I have given so far are the "basic" ones and apply to any "beam". In a moment or two, I will simplify them specifically for use with springs, but first we must tackle the more serious problem of proportioning springs to carry specific loads and at the same time provide sufficient resilience. So long as the load is small, it is possible to design a simple spring without difficulty, but as soon as the loads get high the beamtype spring has to be so deep that it doesn't deflect at all! Or not measurably. Let us look at a typical small tank locomotive, 6-coupled, with each axlebox carrying 8.9 tons, available spring width 4in., and provisional length between hangers of 36in. (The 8.9 ton is the estimated maximum which might come onto each wheel – it doesn't mean that the locomotive weighed 53½ tons – 40-odd would be more likely.) Working it out, at the stress we used for steel before, the single bar-spring would be 4in. wide × just under 2in. deep, and the deflection would be just over ¼in. – 0.279in. to be more exact. This just wouldn't take care of even good rail-joints, let along those met with in the average shunting yard. We have to devise some means of increasing the deflection without raising the stress.

Looking at Formula (10) there is little we can do with "L"; to increase this enough to give the effect we want would make the spring as long as the locomotive itself – perhaps longer. The only other variable we can alter is "I". This, you will remember, is $BD^3/12$ and if we reduce D and increase B (because we must keep the stress within limits), this might do the trick. So it does.

If we make the thickness only ½in., and use a spring 56in. wide the stress will be the same, but the deflection will go up to about 1¼in. – much more hopeful. If, that is, we could find somewhere on the locomotive to fix six springs as wide as the rail-gauge! We can't of course, but we CAN cut it up into slices and set these slices one above the other to make a *laminated spring*.

The Tapered Leaf Spring
If you look back at Fig. 31 again, you will see that the bending moment increases from the ends to the middle – it is greatest under the load. Similar conditions apply to the other forms of loading. This means that the *stress* in the spring also varies along its length. This is a waste of expensive material, and structural beams are normally designed to give a more uniform distribution of stresses. We can do the same with the spring, by shortening the laminations progressively as in Fig. 35, which shows the classical form of the "Half-elliptical" spring; a similar arrangement can be made with the cantilever or "quarter-elliptical". The effect is more or less as shown in Fig. 36(C) – the wide, thin spring we discussed a little earlier is now in the shape of a diamond.

At (A), we have the laminated spring with 5 leaves, shown at (B) looking at the underside of the spring. The diagram at (C) is the flat plate simulation. We cannot use a pure diamond shape, as there must be a minimum width at the ends to carry the shear load, so plate 1 has a width "b". Plates 2, 3, 4, and 5 are the same width, but appear in diagram C on either side of No. 1 *at half the width*. Note that we *could* approach the diamond shape over these other leaves by tapering the ends as shown dotted in the plan view "B". This is sometimes the case, but more usually the ends of these leaves are but slightly tapered in width but with the projecting parts also tapered a little in thickness. With this arrangement, we can set up the locomotive spring mentioned earlier, by using 14 leaves, the lengths varying from 36in. down to 5in. long each ½in. thick, total depth 7in.

In analyzing, the stresses and deflection of such a laminated spring, we have to make some assumptions and conditions. First, that the curvature of each leaf is the same *when unloaded,* so that before clamping up at the center (or end of a cantilever type) the leaves touch

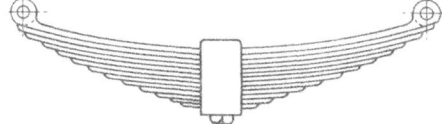

Fig. 35

only at the ends. Second, that the length of each leaf is such that it fits as closely as may be to the diamond pattern of Fig. 86. Third, that there is no friction between the leaves. None of these is met exactly, but the error is relatively small. Taking the semi-elliptic spring of Fig. 31 as an example, the expressions work out as follows:-

Stress $f = \dfrac{3WL}{2n.b.t^2}$ lbf/Sq.in.   (12)

Deflections $\S = \dfrac{3WL^3}{8E.n.b.t^3}$ in.   (13)

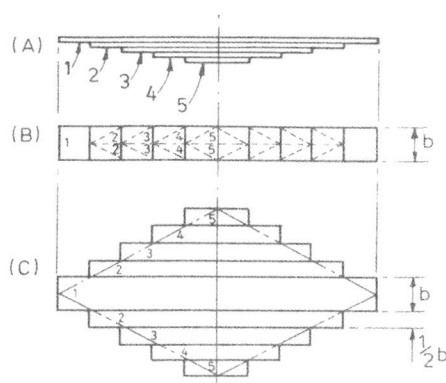

Fig. 36

Where "n" is the number of leaves, "b" is their breadth, in., and "t" the leaf thickness, also inches. (Use mm if working metric.) All other symbols as before. The similarity between equations (12) and (13) with those of (7) and (10) is apparent when it is remembered that $Z = BD^2/6$ and $I = BD^3/12$, but note that "t" is now the *leaf* thickness, not the depth of the spring as a whole.

CHAPTER 7

Leaf Spring Design

Spring Rate

When we were dealing with coil springs, we found that it was convenient to use this conception — load/deflection, measured in lbf/in. or Newton/mm — and there are some cases where this applies also to leaf springs. If you look again at equation (12), you will see that we can rearrange this to give load, W, in terms of the stress, f, thus;-

$$\text{Load } W = \frac{2f.n.b.t^2}{3L} \text{ lbf} \qquad (14)$$

and we can also re-arrange equation (13) to read:-

$$\text{Rate} = \frac{W}{\S} = \frac{8n.E b t^3}{3L^3} \text{ lbf/in.} \qquad (15)$$

Equation (14) is more convenient as we usually design to a fixed stress, and (15) helps when we need a specified "springiness". These formulas for leaf springs are summarized in Table IV opposite.

This table shows, in the first column, alternative formulas connecting load and stress; in the second, the expression giving the shear stress at the ends of the spring; the next column gives alternative formulas for deflection. The fourth column gives the rate, W/§. The final columns refer to charts which I shall be giving you in a moment. You will note that each type of spring requires a different expression, but if you think for a moment, you will appreciate that "nb" in the laminated type corresponds to plain "b" for a solid one.

Using the Formulas

It requires only a casual glance to realize that these formulas, though easy enough to work out (even without a pocket calculator) are not at all easy to use to *design* a spring. We may know the load "W", and choose a stress, "f", but this still leaves no less than *four* other dimensions to work out. Even if we know the desired deflection (see Col. 3) and "f" and "E" are decided, there are two unknowns to calculate. Those who got "A"s in math – and perhaps even the odd mathematics professor – will know that you can solve for ANY number of unknowns, provided you have an equal number of equations, but here we don't have that situation. It is the same "catch 22" problem which faced the bridge designer, as I mentioned earlier on. So, to ease matters for you I have prepared

TABLE IV

In all cases:
- W — max load, lbf.
- L = spring length–see diagrams. Inches.
- b = breadth of spring or leaf. Inches.
- t = depth of spring or *thickness of leaf*. Inches.
- n = number of leaves.
- E = Young's Modulus of Elasticity, lbf/sq.in.
- f = Bending stress, lbf/sq.in.

In metric units; load W in Newtons, all dimensions in millimeters. Young's Modulus in N/sq.mm, gives stresses in N/sq.mm and deflection in mm.

Spring type	Max. Bending Stress lbf/sq.in.	Max. Shear Stress lbf/sq.in.	Max. Deflection Inches	"Rate" of Spring lbf/in.	Values of K Fig. 37	Values of δ Fig. 39
Solid Cantilever	$f = \dfrac{6WL}{bt^2}$ $\left(W = \dfrac{fbt^2}{6L}\right)$	$\dfrac{W}{bt}$	$\dfrac{4WL^3}{Ebt^3}$ or $\dfrac{2}{3} \times \dfrac{fL^2}{Et}$	$\dfrac{1}{4} \times \dfrac{Ebt^3}{L^3}$	$\times 4$ for $n=1$	$\times \dfrac{8}{3}$
Laminated Cantilever (Quarter Elliptic)	$f = \dfrac{6WL}{nbt^2}$ $\left(W = \dfrac{fnbt^2}{6L}\right)$	$\dfrac{W}{bt}$	$\dfrac{6WL^3}{Enbt^3}$ or $\dfrac{fL^2}{Et}$	$\dfrac{n}{6} \times \dfrac{Ebt^3}{L^3}$	$\times 4$	$\times 4$
Solid Simply Supported	$f = \dfrac{3WL}{2bt^2}$ $\left(W = \dfrac{2fbt^2}{3L}\right)$	$\dfrac{W}{2bt}$	$\dfrac{WL^3}{4Ebt^3}$ or $\dfrac{1}{6} \times \dfrac{fL^2}{Et}$	$4 \times \dfrac{Ebt^3}{L^3}$	As read for $n=1$	$\times \dfrac{2}{3}$
Laminated Simply Supported (Half-elliptical)	$f = \dfrac{3WL}{2nbt^2}$ $\left(W = \dfrac{2fnbt^2}{3L}\right)$	$\dfrac{W}{2bt}$	$\dfrac{3WL^3}{8Enbt^3}$ or $\dfrac{1}{4} \times \dfrac{fL^2}{Et}$	$\dfrac{8n}{3} \times \dfrac{Ebt^3}{L^3}$	As read	As read

some charts.

These will give you an *approximation* to the spring dimensions, and it is then very easy to refine the design using one or other of the formulas. A much quicker (and easier) procedure than the usual "trial and error". Note that I say "approximate" solution. Absolute exactitude is not possible for the initial figures found from the charts. First, because the range of springs is large – Fig. 37 runs from a spring of 1 in. span carrying a load of 1lbf (or 1ft. long carrying only 1½OZ.!) to one which would stand 1cwt. on a span of 9in. Second, because in order to prepare them, I have had to "manufacture" some special graph paper. This has been photocopied, and before you see it, there will also be a tracing operation and further photo-reproduction to get it into print. Some slight distortion will be inevitable. However, the result will be a good first approximation, and you should land up at least within one gauge number.

The charts are drawn for a semi-elliptical laminated spring – the last one on Table IV – and for carbon spring steel stressed at 70,000lbf/sq.in., the value of "E" (Young's Modulus) being 29,000,000. Conversion to other materials or stresses is very easy, and I will deal with that when the time comes. I have also made three other assumptions: (i) that you know the load; (ii) that the span, "L", is fixed by the geometry of the model you are building. This is usually more or less the case though it isn't a big job to alter it if you find that "it just won't do after all". The third assumption needs a little more discussion, and applies to the deflection.

Flat springs fall into three broad types: (a) Those where the deflection is unimportant – a spring controlling a ratchet is an example. (b) Those for railway locomotives and rolling stock Here the main function of the spring is to distribute the load equally between the driving wheels, bogie wheels, etc. *Some* resilience is required to accommodate slight track irregularities, but these are (or should be) relatively small. (c) Those for road vehicles. These, again, are required to distribute the load between wheels with the vehicle level, but also have to accommodate considerable irregularities.

In case (a), it is only necessary to check that the spring will deflect sufficiently to allow the ratchet or whatever to work. In case (b), the calculation of deflection is needed only to determine the initial curvature of the spring so that the loaded configuration is as it appears on the prototype – in the majority of cases, the loaded position will be with the top leaf almost flat. In case (c), I have again assumed that the spring will be flat when normally loaded, but **you** will have to decide on the desired "rate" (column 4) to get the riding qualities needed. This is done by adjusting the values of "n", the number of leaves, and "t", their thickness.

Designing with the Charts
When using Chart 38, take care to read the scales of "b" on the diagonal lines correctly; thickness increases from right to left.
The initial work is done in two stages. You know the load, and have decided on the span, L, so can calculate W × L. Look at the first chart, Fig. 37. Suppose W = 8lbf and L = 2½in.; then WL = 20. From 20 on the horizontal scale read upwards. You can use one leaf, when K = 72, or 4 leaves, K = 18, or 8 leaves, K = 9. Now turn to Fig. 38. From the left-hand vertical scale read off these values of K.

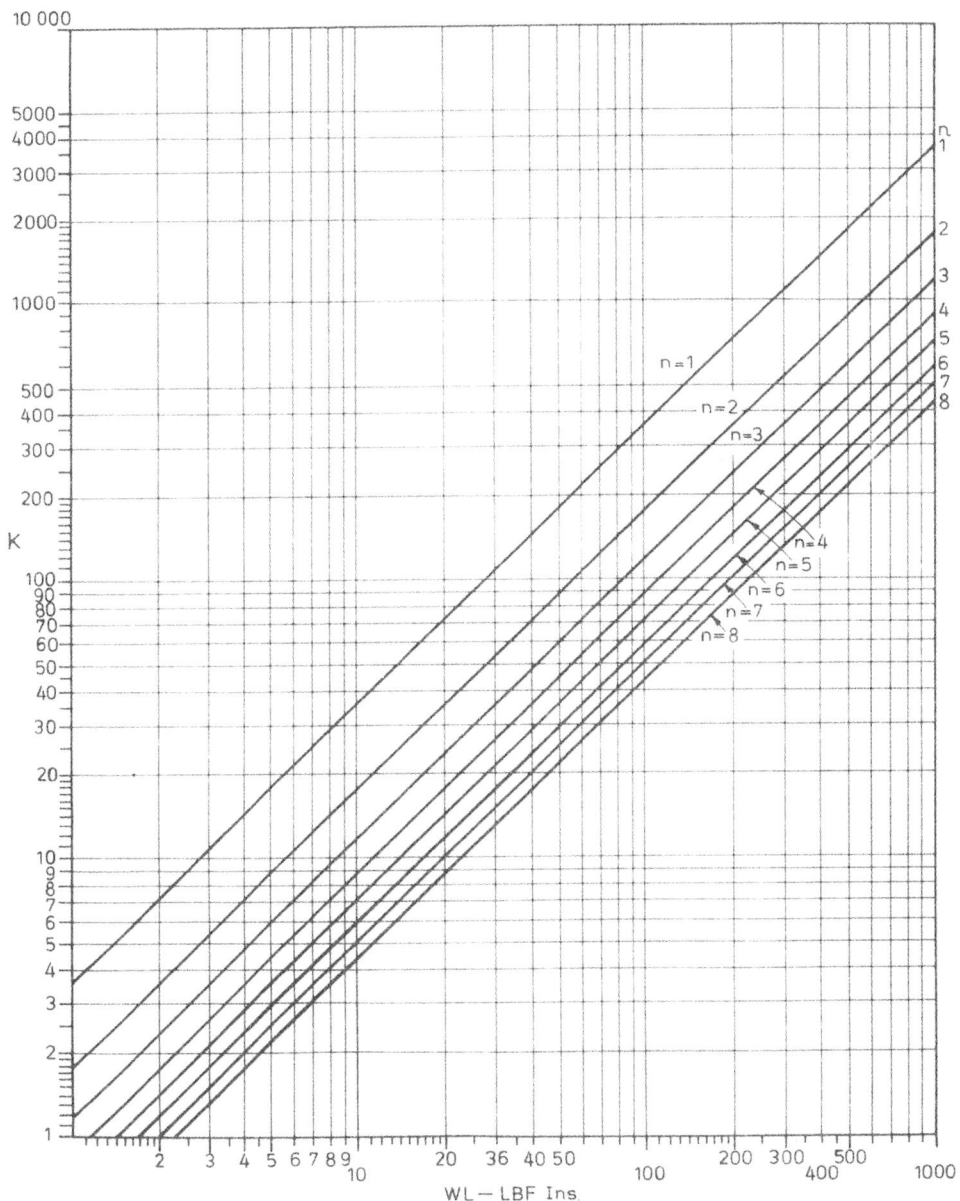

Fig. 37

Leaf Spring Design

(The five sloping scale lines are for different values of the breadth, "b".) For K = 72, you can use a leaf 0.028in. thick × ½in. wide, or one 0.047in. thick × 0.2in. wide and so on. For K = 9, you have the choice of between 0.012 × 0.4in. wide to 0.023 × 0.1 in. wide. In practice, of course, you would use the s.w.g. numbers I have shown at the top of the chart and interpolate between the sloping reference lines to get a reasonable width. For the 8 leaf, you might use 28 s.w.g. material ¼in. wide.

Now turn to Fig. 39. Here the sloping lines are for the usual s.w.g. numbers. The span is 2.5in., so read upwards from this to meet the 28 s.w.g. line and read off the deflection from the vertical scale. It comes out at 0.25in. – not unreasonable. All you have to do now is the check that this is O.K. Put the now known values of W, L, n, b and t into the formula for stress (Table IV, page 55), and you will find that it comes out at 66,700lbf/sq.in., which is satisfactory. Put THIS figure into the deflection formula in the same table, and you get § = 0.2396 (about 0.24)in., which is not far off the original estimate. In fact, on my original chart the intersection is just below the ¼in. mark.

Now suppose we want to use the chart for Brass. The safe working stress is now 28,000 (see Table III, page 51) and E = 16,000,000. Read off the value of K as before, but then multiply this by the ratio of the working stresses, 70/28 = 2.5. Instead of K = 9, you now use K = 9 × 2.5 = 22.5, when using Fig. 38. You would now need either about a 23's gauge leaf ¼in. wide, or one ⅜sin. × 24's G. Transfer to Fig. 39. From L = 2.5, read up to the 24's gauge line to find that § = 0.17in. This must now be adjusted by two factors to allow for the different stress and the different value of E as

the chart was drawn for f = 70,000 and E = 29,000,000. (Look at the formulas in column 3 of the table. The lower stress will reduce the deflection, and the lower value of E will increase it.) Hence:

§ (Brass) = 0.17(16/70) × (29/16)
= 0.07in.

Though brass is the weaker material, the deflection is reduced, *mainly because we have to use a lower stress.* This first increases the thickness of the leaves (see the formula for W, column 1) and then has a direct effect as shown in the alternative formula in column 3.

Other Forms of Spring

These charts are drawn for the commonest type of spring, the semi-elliptical laminated, shown at the bottom of Table IV. The 5th and 6th columns show the correction to be made for the other forms. Thus, for a single-bar spring use n = 1 – a fairly obvious adjustment! For a laminated cantilever (or ¼-elliptic) spring, the value of K should be multiplied by 4. There is no correction needed for Fig. 38 for TYPE of spring so long as "L" is measured as shown in Table IV, but the deflection determined from Fig. 38 should be multiplied by the factors shown in column 6 of Table IV.

There is one circumstance where further adjustment may be needed, and that is if the *figures go off the chart.* On Fig. 37, you might have a load of 200lbf on a 10in. span – WL = 2,000. However, K is proportional to WL, so that all you need do is to read off for WL = 200 and multiply by ten. Similarly for Fig. 39, where deflection is proportional to L squared. Suppose you have a span again of 10in.; read off the deflection for L = 5 and multiply by $(10/5)^2$ – i.e. by 4. There is little risk of running off-scale on Fig. 38!

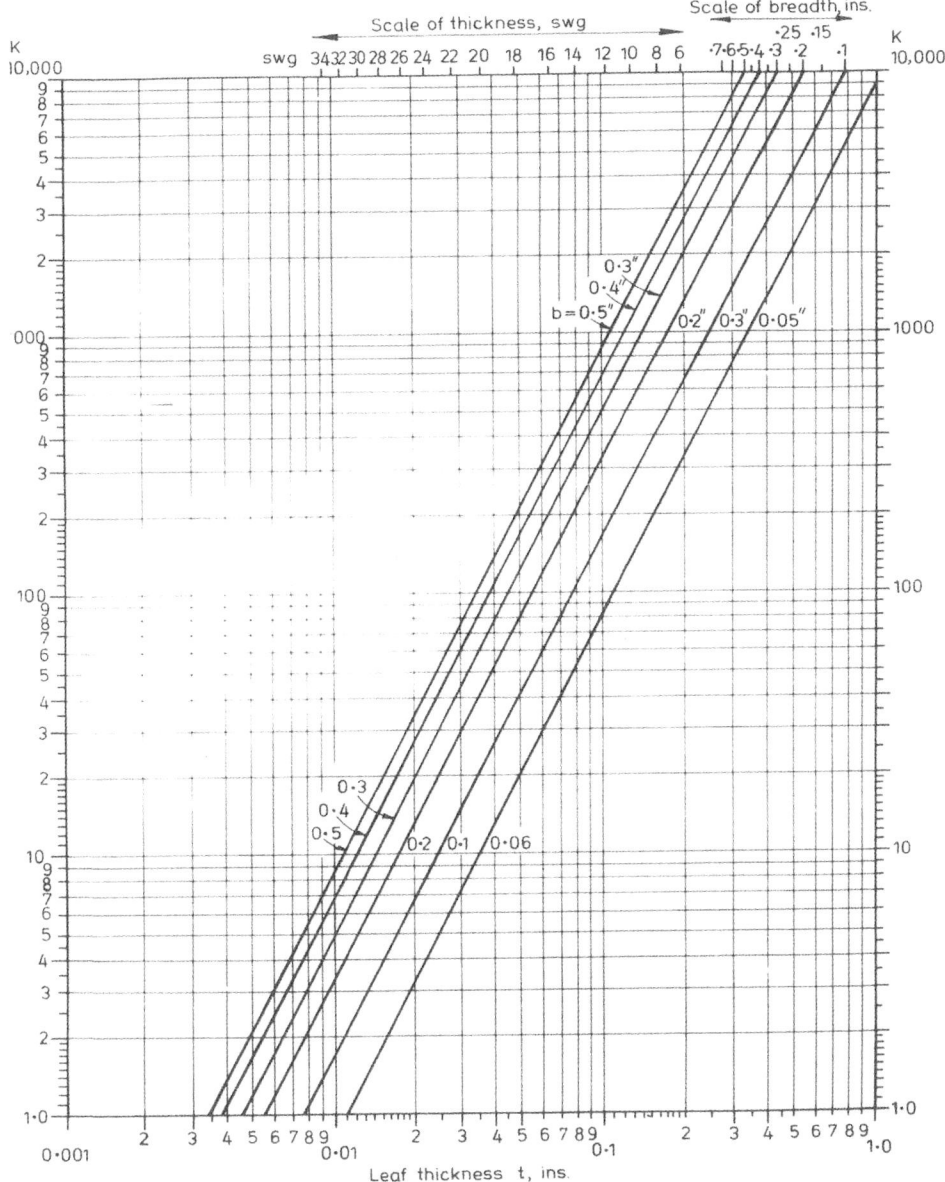

Fig. 38

Leaf Spring Design 59

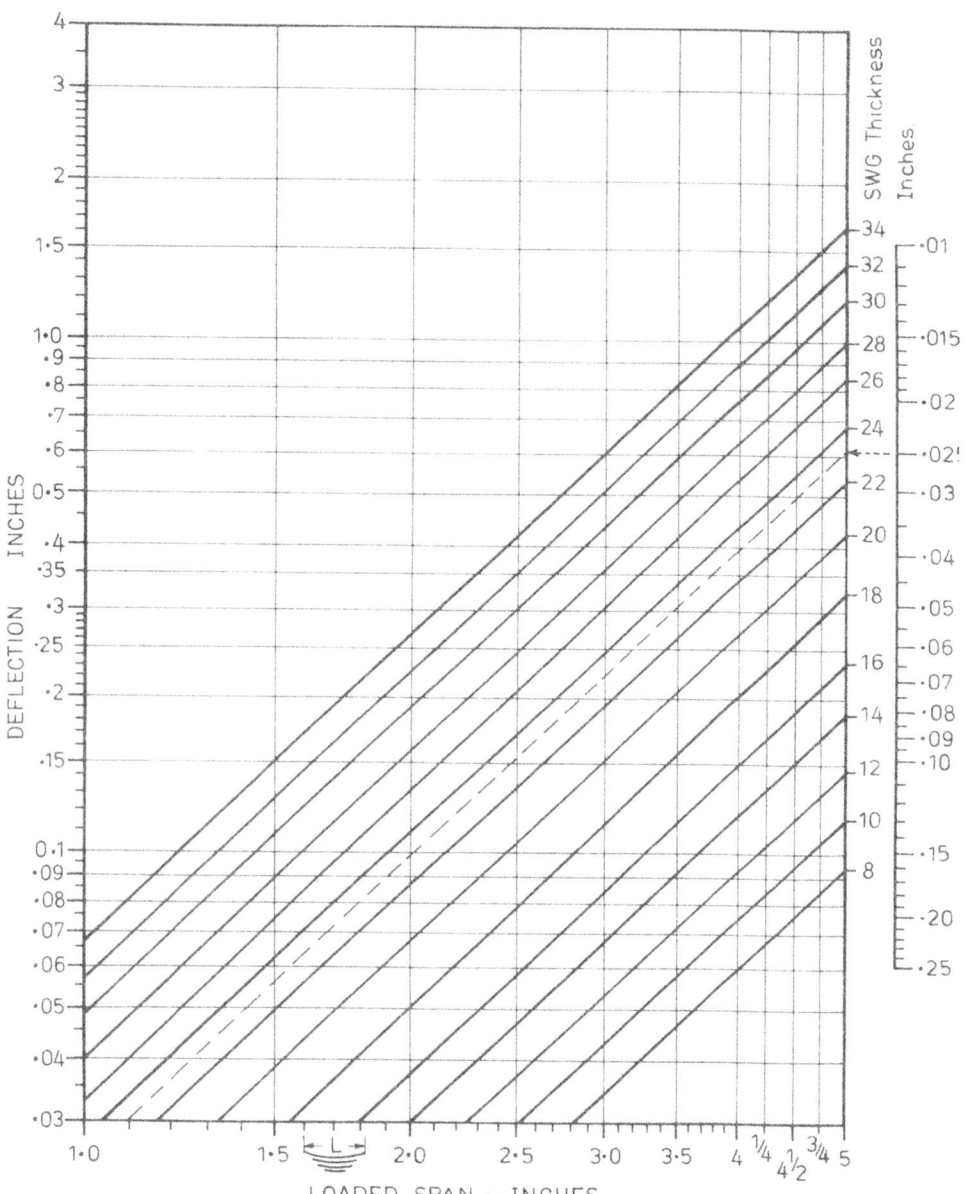

Fig. 39

60 SPRING DESIGN AND MANUFACTURE FOR HOME MACHINISTS

The Clamped Spring

I have had this method of evaluating springs questioned in the past because textbooks which readers have referred treat the *half-spring* only, and use the reaction at the support instead of the load at the center – in other words, they treat the semi-eiliptical as if it were two quarter elliptical springs. Now, this makes NO difference to the end result – the difference between their formulas and mine takes account of the different treatment. See Fig. 40. At (a) is the textbook treatment. Now, the reason for this is that in *full-size* practice the center buckle is a hefty affair, and really does clamp the spring so that it behaves like two cantilevers from the center. But note that in this case *the span is NOT the distance between the spring anchorages*. In most MODEL work, this center spring buckle is far too weak relative to the spring leaves to have any effect at all. We can happily and safely assume that the load W is applied directly to the center of the span as shown in Table IV. However, on larger scale models – for example, above (and possibly at) 1/8 scale locomotives, and quarter size road engines – then the scantlings of this center buckle may be sufficient to have an effect. In that case, you can still use my charts and Table IV, but the value of L is then determined by measuring it as shown in Fig. 40b.

Important Note

Whichever method is used, it is vital that the effective spring length be *that which applies when the spring is loaded*. That is, when the stress is at a maximum. With some old-style road vehicles, the initial camber on the spring is so great that there is a considerable difference between the loaded and unloaded span between the end spring-shackles.

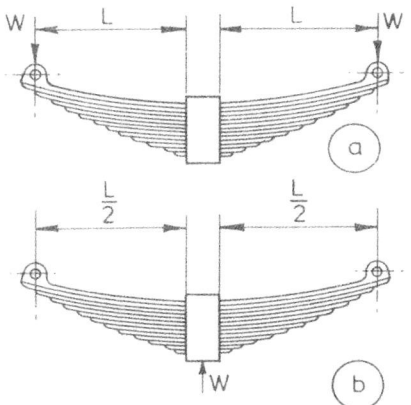

Fig. 40

Scale Leaf Springs

A quick glance at Table IV will show that it is just not possible to fit a spring to a model in which the various dimensions are simply reduced to scale size and hope that it will act in a "scale" way. This is true even if the *weight* of the model behaved "scalewise". Ideally, a model at 1/10 scale would weigh $1/10^3 = 1/1000$ of the weight of the prototype, but you have only to look at the thickness of (e.g.) cylinder castings to realize that this is seldom the case. If the spring is scale size in every respect, it cannot function properly. Unfortunately, we have no "scale material", and even if we had, we should need a different metal for each scale. The solution can be (a) to use dummy springs where they show and conceal operative springs somewhere else. This is perhaps the best solution for working models, where fidelity to prototype is not the aim. (b) To use a compound spring, where most of the load is carried on steel leaves but the rest on very much weaker material. Tufnol is the usual choice. (c) To use leaves made of a number of "leaflets" of very thin

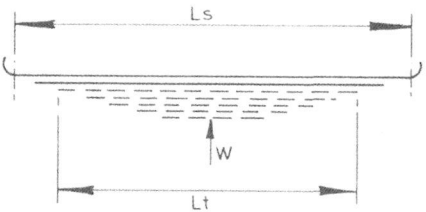

Fig. 41

metal. (d) To weaken most of the leaves, for example by milling a slot down the center. We will look at these in turn.

The Compound Spring

Fig. 41 shows the arrangement, where the full lines represent the strong material (steel, bronze, or whatever) and the dotted lines the weak, usually Tufnol. The behavior of a spring is rather complex. When first introduced, most people regarded the Tufnol as mere packing, but it can and does contribute to the load bearing capacity and affects the deflection. In effect, we have a diamond-shaped spring like Fig. 42, in which the shaded area represents the steel, bonded full strength to the unshaded Tufnol area. This means that in practice the two deflect together, and we can use this fact to arrive at the proportion of the load carried by each. Look again at Fig. 41; in effect we have two springs, one above the other. At first sight, the vertical deflections (relative to the ends) measured at the center will be different, but you will recall that the TRUE deflection so far as stress is concerned is the change in radius of curvature. This change will be identical for all the leaves in both parts of the spring so long as they act together, and this they are bound to do.

Remember the basic equation we looked at some time ago on page 50 which reads:

$$\frac{f}{\frac{1}{2}D} = \frac{M}{I} = \frac{E}{R}$$

(D = t = thickness of leaf.)

As D (leaf thickness) and R are the same for both steel and Tufnol, we can write:

$f_s/f_t = E_s/E_t = 29M/1.5M = 25.2$ about 25.

as, from Table III, the ratio of E for steel and Tufnol is 29/1.5.

This does not mean that we can write that the stress in Tufnol will always be 70,000/25 if we use 70,000 in the steel leaf; the *actual* stresses will depend on several other factors in a compound spring, but the *ratio* of stresses will be 1/25. We can use the same reasoning to ascertain the stress ratio for any other combination of materials.

We can now use this ratio to find the distribution of LOAD between the two parts of the spring. Look at Eqn. (14) – shown also in Table IV:

$$W = \frac{2}{3} \frac{fnbt^2}{L}$$

from which

$$\frac{W_s}{W_t} = \frac{f_s n_s}{f_t n_t} \times \frac{L_t}{L_s}$$

(b and t are the same for both materials).

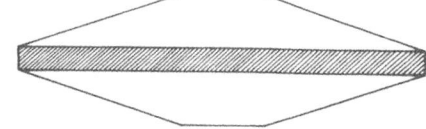

Fig. 42

as $f_s/f_t = 25$
$$W_s = 25W_t(L_t/L_s \times n_s/n_t)$$
where L_t and L_s are as in Fig. 41
n_s is the number of steel leaves
n_t is the number of Tufnol leaves
or
$$W_t = 0.04W_s(L_s/L_t \times n_t/n_s)$$

The total load, $W = W_t + W_s$, hence

$$W = W_s \left(1 + \frac{1}{25} \times \frac{L_s n_t}{L_t n_s}\right) \quad (16)$$

From the prototype you can ascertain the values of Ls and Lt and the number of leaves of each by assuming (in the first instance) that the top two leaves will be of steel (see Fig. 41) and then work out how much load will fall on the steel leaves. Then apply Fig. 37 chart (or the formula in Table IV) for n = 2 and design the spring as previously explained. You *may* find that you have to do it twice, inserting new values, for it can happen that two leaves will give unreasonable thicknesses, and you will have to revert to a single steel leaf. It is not often that you will find that you need more than two.

If, however, you have no prototype to refer to you must take a few design assumptions and then "try it and see". Sketch out a spring which "looks right" for the job in hand. I find that Ls/Lt usually lies between 1.07 and 1.21, and so start with for example, 1.125. Work out Eqn. (16) for perhaps 7 leaves in all, and alternatively for one and two steel leaves. It may come out something like

$$W = 1.3W_s$$

You can then follow one of two procedures.

(a) Use the charts Figs. 37 and 38 for ns = 1.3 leaves – just interpolate between the lines marked "1" and "2" – and determine the breadth and thickness in the usual way to carry the FULL load W multiplied by the length of the top leaf. Then sketch out the spring, putting in the "number of leaves you first thought of", check the value of Lt, and see how it works out.

Or (b) Calculate the value of WL for the value of W_s falling on the steel leaves alone – as determined from Eqn. (16) – and read from the line which gives "n" as the actual number of steel leaves. Both methods will give the same result. (I give an example shortly.)

If the leaves come out too thick (or too broad) try with two steel leaves – or even more in extreme cases. If too thin or too narrow and you are already down to a single steel leaf then, if the excess is small, use fewer of Tufnol. However, if the discrepancy is large, you will have to change the material of the master leaf, the idea being to find a smaller ratio of E/f (Young's Modulus divided by working stress). Unfortunately, the choice is not wide, the ratios being as follows (but work it out for yourself as well!):

Steel/Tufnol	25/1
Phos. Bronze/Tufnol	14/1
Be. Copper/Tufnol	15.6/1
Brass/Tufnol	14/1

Note that you cannot use Steel/Brass, as in such an arrangement the brass would be overstressed. (I will refer to the use of plain mild steel in a moment.)

In the absence of elaborate computer programs, the design of these composite springs is, at best, a messy business, not at all easy, and you must be prepared for several goes of "suck it and see". Keep an eye on the expressions in Table IV and juggle with the dimensions which seem likely to give the best effect – and don't be afraid to alter even the actual length of the spring if the design will allow this. It helps. There is also the difficulty that I have *had* to oversimplify several matters, and in any case, Tufnol is somewhat variable, especially in the value of "E". For what it is worth, I find that over the normal or typical range of numbers of leaves the Tufnol part behaves more or less like 0.3 steel leaves, but in extreme cases, this can be as little as 14% and as great as 65%. This being so it is only prudent to *make up a trial spring* after the calculations have given you a guide, and test it on the bench. Once you have determined the "equivalent steel spring" you can use chart Fig. 39 to estimate the deflection.

Example
A "near scale" spring is required to carry a load of 12½lbf on a span of 3.75in., having 8 leaves as near as possible to 0.058in. thick and 0.32in. wide. Calculate the number of leaves of Tufnol and the initial deflection if it is to lie flat under load.

The nearest standard thickness is 1/16in., so use this.
For steel/Tufnol, stress ratio $E_s/E_t = 25.2$. To estimate L_s/L_t, draw a diagram as Fig. 43 and measure L_t as shown. From which $L_t = 3.4$, and $L_s/L_t = 1.1$. (Note that I have assumed the length of the bottom leaf to be 1in. – this is a matter of choice.)

From Eq. (16):-

$$W = W_s \left(1 + \frac{1}{25.2} \times \frac{3.75}{3.4} \times 7 \right)$$

assuming 1 steel and 7 Tufnol.

$$W = W_s \times 1.306$$

Hence $W_s = 12.5/1.306 = 9.56$
(Wt = 2.94)

And $W_s L_s = 9.56 \times 3.75 = 35.85$

From Fig. 37, K = 132

From Fig. 38, reading from K = 132 gives: $\underline{t = 0.063}$ and $\underline{b = 0.22in.}$ This is too narrow.

To use the same value of K and a breadth b of about 0.32, we would have to use t = 0.052in. This is very nearly 17 s.w.g. and when checking from Fig. 39, we get a deflection of 0.155. This is rather small, and in any case, it would be difficult to obtain Tufnol to match the thickness. (The spring would attract adverse comment if all leaves were not the same, though very small differences might not be noticed.)

So, let us try **phosphor bronze,** which has a lower value of Young's Modulus (E) for the top leaf. And let us work this at a rather lower stress, hoping that thicker leaves will not reduce the deflection too much.
The stress ratio is now 14, the other figures the same.

From Eqn. (16) $W = 1.55 W_b$
(W_b = Load on bronze leaf)
$W_b = 8.1lbf$ and $W_b L = 30.2$.

Fig. 37 gives K = 110, but this must be corrected for a lower working stress, for example, 40,000. So, $K = 110 \times 70/40 = 193$.

Fig. 38, using this value, gives $\underline{b = 0.3}$ with $\underline{t = 0.0625.}$

This is better. Check for the deflection. It is easier to use the simpler of the two formulas for deflection on Table III, rather than to use Fig. 39 and correct the reading for both stress and Young's Modulus.

Deflection, $\S = \dfrac{fl^2}{4Et}$

$= \dfrac{40,000 \times 3.75^2}{4 \times 15.5M = 0.62}$

$= 0.147$ in.

This is a reasonable figure, and though the leaves are slightly thicker than originally asked for, the width is very close. It remains only to check the stress in the Tufnol to see that it is safe.

$ft = f_b/14$
$= 40,000/14$
$= 2860 lbf/sq.in.$

This is safe.

The calculation shows that 8 leaves are needed, one phosphor bronze and 7 of Tufnol all 0.3in. wide and 1/16in. thick. Check it yourself.

To ease manufacture, I would make this spring of 5/16in. wide leaves (to ease the machining of the center buckle) with 1 Phosphor Bronze and 7 Tufnol leaves, all 1/16in. thick, and give perhaps 3/16in. initial "set" to ensure that the top leaf was just slightly curved under load. But it would be essential to make a trial assembly first, to check both load and deflection.

The "Leaflet" Spring

To take the previous example, with this type we should *apparently* have only 8 leaves, but each would be made up of a number of much thinner leaflets or blades. This method was, I believe, first suggested by the great "LBSC" many years ago. It "works", but has a disadvantage which is best illustrated by an example. Let us work out the previous spring, still using 8 leaves but with each leaf made up from *five* blades all of steel.

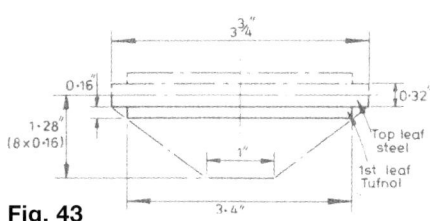

Fig. 43

Ideally, their thickness would be 0.058/5 = 0.0114in. The nearest standard gauge is No. 32, at 0.0108in. – near enough, as it is only 0.0006in. thicker. I am going to work this one out "backwards", so that you can see this method.

From Fig. 38, "K" for 32 s.w.g. × 0.32in. wide = 6

Look at Fig. 37. There is no reference line for a total of 40 leaves, so use that for four, and multiply K by ten.

When K = 60, and n = 4, *WL reads 68.* But the required value of WL is 47; if we apply only WL = 47 to a spring which accepts 68lbf/in. at a stress of 70,000lbf/sq.in., this means that the *actual* stress will be lower in proportion.

Hence, f = 70,000 × 47/68
$= 48,300 lbf/sq.in.$

This also means that the spring has a considerable overload capacity, thus:

Max. safe load = 12.5 × 70,000/48.300
$= 18 lbf$

Now check the deflection from Fig. 39. § for 3.75in. span at 32 s.w.g. reads *0.73in.* But this must, again, be corrected, as the chart is prepared for stresses of 70,000lbf/sq.in. Thus:
Actual $\S$ = 0.73 × 48,300/70,000
$= 0.504in., about ½in.$

Leaf Spring Design 65

The last figure illustrates the disadvantage – the spring is far too "lively" for its application as a locomotive axlebox spring. The chassis would bounce up and down at the slightest change in load – there might even be a difference in running height between "bottom glass" and "full glass" on the water-gauge! There is a visual problem, too; in service the small "leaflets" may well work slightly sideways, giving the game away – even if we bevel the top sides of the top leaflets and bottom sides of the bottom ones in each group, to give the impression of eight solid leaves. The other problem is that we are both stretching the "mathematical model" – the theory – to its limit and, in addition, we now have 39 friction surfaces, instead of the seven in the proper 8-leaf spring. Both of these factors lead to the probability that the actual spring may depart more than usual from the calculations. We shall be looking at this in connection with all leaf spring design in a few moments. Despite all this, the "leaflet spring" is often a practical alternative to the compound type.

The Slotted Spring

This is another method of bringing a spring of scale appearance down to scale performance. See Fig. 44. The top leaf and the short bottom leaf are plain, but all the others are slotted as shown. This slot is carried almost to the ends, leaving an unslotted length equal at least to the full width of the leaf. The spring now behaves as if it were made up mainly of leaves only 2b wide, the top leaf counting as B/2b in number. (The very short bottom leaf can usually be ignored.) Suppose we have 7 leaves 0.4in. wide (B) six of them slotted ¼in. wide. Then the width "b" to use on the chart or formula is 0.4 – 0.25 = 0.15in. B/b = 2.67, so that the number of leaves, n, will be 8.67, about 8.7. (Six are 0.15 effective width, and the master leaf is equivalent to 2.7 more.) This method is effective, but it is not easy to machine hardened and tempered spring steel. In addition, there is a risk of the narrow part of the leaves splaying inwards or outwards. For this reason, some users arrange the center as shown in Fig. 44b. It is usual, too, to set an extra, very short, unslotted leaf at the bottom of the stack.

Mild Steel Springs

You can, of course, use any material you like. There is an erroneous impression that spring material must be *hard*. This is wrong – remember your catapult elastic? What is needed is a high ratio of working stress to Young's Modulus – f/E – if the spring is to be economic and effective. The apparent hardness is a consequence of the metallurgical or mechanical treatment accorded to the metal to achieve this high ratio. Mild steel *can* be used, but the working stress must, of course, lie below the elastic limit (or yield point) which is relatively low; at the same time, "E" is higher than that of proper spring steel, at about 30×10^6. The working stress should not normally be allowed to exceed about 25,000lbf/sq.in. for active springs – perhaps 30,000 for

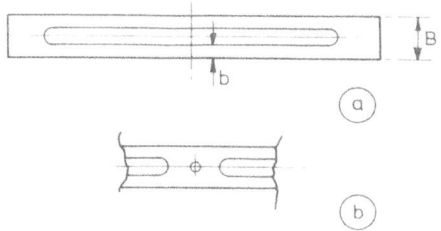

Fig. 44

springs which just have to "sit and bear" the load. Though I have made successful laminated springs for light work using *shim steel,* which appears to be happy at up to 30,000lbf/ sq.in., probably due to the cold rolling operation used in its manufacture. For general use, mild steel springs will be heavier and have more leaves than those of spring steel.

Design Uncertainties

I have used the term "uncertainty" rather than "error" or "limitation" because the differences between the calculated results and those from tests are, as a rule, due to just that. If you had sufficient computer power, and were adept at finite element analysis – and had specialized in spring work for your Ph.D. – you could get very close indeed! But for *our* purposes the method will serve its office – to get a reasonable first approximation to the dimensions of the spring. But it may not hurt to look at some of the sources of uncertainty.

First, *friction.* In service, each leaf works against its two neighbors, sliding lengthways. The deflecting force will be resisted in some small degree by this friction. The result is that with slow application of load the deflection will be reduced. Further, when the load is removed – again if done slowly – the spring will not return to its original position. Neither of these phenomena will be noticeable if the load is applied repeatedly, so that it is not important so far as desired deflection for a suspension spring is concerned. Indeed, this friction can be an advantage, for it acts as a *damper,* reducing the tendency of the spring to oscillate under suddenly applied loads. (In fact, one car which I ran some 40-odd years ago, but ten years old even then, actually had layers of friction material set between the leaves, to help the shock absorbers!) However, it *is* desirable that the surface finish of the faces of the leaves should be good, and the leaves occasionally lubricated, to reduce friction if possible.

Second, *the basic assumptions.* We have used what is called the "simple theory of bending" in analyzing the behavior of these leaf springs; and one of its basic assumptions is that the "beam" is virtually straight and that deflections are small. Once the spring is loaded – and properly, the top leaf should be almost flat at normal load –we are not far out, but in the initial stages of the deflection, the "beam" is very much curved, and the "simple theory" is not really applicable. (We can, of course, use the theory applicable to curved beams instead, but this would make things very much more complicated!) It is sufficient to note that the "simple theory" used here is prone to *overestimate* the deflection by perhaps 5% or so. The stresses also will be slightly higher, but not sufficiently so to cause any alarm.

Third, *loading on the leaves* is far less simple than is implied in the analogy of a diamond-shaped single leaf. In particular, the transfer of load from one leaf to the next is much more complicated than we have assumed. This means that it is NOT valid practice to test a single leaf in isolation for load and deflection and then say that "n" leaves assembled together will carry "n" times the load at the same deflection. The theory is not too far out *for the spring as a whole,* but it will fall down if you try to take it apart! You must carry out your tests on mock-up assemblies of the whole spring. The effect here may be 5% either way, depending on the number and proportions of the spring leaves themselves.

Finally *tolerances.* If you look at Table

IV (page 55), you will see that in every column we are multiplying several factors together, and that in some cases these factors are "squared" or even "cubed". This means that tolerances on the actual spring material can have an effect. If you are 1 % out in both thickness and width of the leaves, you may be 3% out in the load or stress calculation, and a little over 4% out in the deflection. If you are in addition, 1% out in estimating the *loaded* value of "L", *and* misjudge the load as well, you can be way, way out!

Fortunately, these tolerances can work in opposite directions, and can, in total, cancel out the deviations caused by the other factors I have mentioned. But they all add up to one basic principle – *you must make a practical test.* The whole object of this book, as I said at the beginning, is not to enable you to do without such checks, but rather to reduce the number you have to make, by giving you at least a reasonable starting point. To comfort you, though I don't make many leaf springs (and very few of the "compound" type), well over 50% of mine come out "near enough" at the first trial.

CHAPTER 8

Making Leaf Springs

I do not propose to go into too much detail here, but will try to cover the points which really matter. The first of these is the MATERIAL. I have already given the stress and Young's Modulus figures. These can vary just a little with small variations in analysis, but not enough to worry about. However, if you "just pick up" some "really good" spring steel you may find yourself a way out with your calculations – and even more in trouble if you try any heat treatment! For all normal purposes, common "carbon spring steel" is good enough. If you need any in large chunks, look around for some old – really old – farm cart or implement springs, or pre-1914 coal wagon springs. These will be fairly thick and can be annealed, machined, and re-heat-treated with no difficulty. Even cold-chisel steel will harden and temper to spring quality if need be. Harden from about 780 deg. C and temper at 310 to 320 deg. C – "Blue", to "Pale Blue", which comes just AFTER "dark blue", by the way. I don't temper springs nowadays, but use the "Austempering" process. This gives much better quality, and provided you have a thermometer which goes up to about 350 deg. C is much more accurate.

Curvature
This should be the same for all leaves – you should NOT make the radius of each leaf a little larger to allow for the thickness of the one above. In fact, it is better to use a slightly *lower* figure (smaller radius). The initial assembled radius must be that which will, under load, bring the spring flat, though you may care to design for it be flat at about 10% overload. (Provided, that is, that the prototype does not call for a curved spring. Those on pony traps had considerable curvature!) The curvature is found easily enough from the deflection, as shown in Fig. 45. To be absolutely accurate, "L" should be the unloaded length, but if you use the same value as for designing, the spring it will

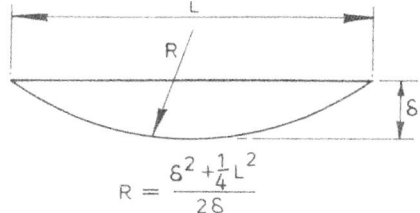

Fig. 45

make little difference. As all leaves are to have the same curvature, it is easiest to run a whole length of leaf material through the bending rolls and then cut off lengths as required.

Tufnol must be curved under heat – 5 to 10 minutes at 120 deg. C in the domestic oven. There is some spring back, and you will have to experiment with the radius of the former or coil. If a metal former is used, leave time for it to get up to temperature.

Leaf Length

The overlap of each leaf above the one below should be determined from a scale drawing of the diamond shape shown in Fig. 36, but I prefer to set limits to the length of the bottom leaf, making it *not less than* three times the width of the center buckle and *not more than* one third of the length of the top leaf. If Fig. 36 gives a bottom leaf within these limits, use it, but if not, modify the lengths accordingly. The overlap should be the same for each pair of adjacent leaves, right through the stack. Wherever possible you should always have *two* top leaves of the same length, the very top one being fitted with the eyes or bosses for the shackles and the next one curved under these eyes a little, as in Fig. 35. This is a safety measure, in case the master leaf breaks. The margin is sufficient as a rule for the spring to avoid catastrophic collapse, but if there is no back-up leaf, this can happen. Springs for road vehicles often had buckles at the ends of the middle leaf. This was, I believe, to ensure that it and the leaves above acted together as a laminated spring on those occasions when a "bounce" left the spring supporting the weight of the axle and wheels.

Shackles and Eyes

The center shackle must be designed to suit the job you have in hand, but you should always have some means of locating the various leaves endways. This is usually done by fitting a small (and it should be small) bolt right through. There have been complaints that spring steel is difficult to drill; not surprising! The trick is to use a short, stub length, drill, very high pressure, low speed – use a hand-drill if need be – and with a backing piece of steel clamped to the leaf to take the thrust as the drill breaks through. Needless to say, the drill should be VERY sharp, and those who have any *carbon* steel drills (as opposed to HSS) of stub length will find little difficulty. (Carbon steel is harder than HSS.) But don't try to make a broken "jobbers" drill into a "shortie", for the upper parts of the fluted length are somewhat softer than the point.

The eyes at the ends of the top leaf do present problems. First, however, their size. This is a case where "scale effects" work in your favor; a scale size pin will be more than ample – the stresses will be lower by the factor 1/scale. However, if you want to work it out you must allow for the fact that the load may NOT be equally shared between the two ends all the time. We usually assume that 5/8 of the center load *may* come onto one end occasionally, and design for that. Secondly, you must add any other loads which may come onto the pin. On a road vehicle, both the accelerating forces and those due to braking will be born by one end of the spring. (They can also put more bending loads in the spring itself, too, but for model work, this is seldom of importance.)

To allow for shear stresses, a mild steel pin should have diameter –
d = 0.007 $\sqrt{W}$ for "live" loads
or d = 0.0058 $\sqrt{W}$ for "dead" loads.
(A "live" load is one which can vary cyclically between 0 and W.) W is the actual load on the pin, i.e. ⅝ the center load. To give adequate area as a bearing:

$$d = \frac{0.0006W}{B}$$

for mild steel or silver steel.

B = bearing length.

$$\text{or } d = \frac{0.0004W}{B}$$

for casehardened steel

I do NOT recommend hardened and tempered silver steel for such pins. Case-hardening gives much better wear resistance and a more ductile core.

These formulas will give stupidly small pins if worked out for model loads, but they can be useful when designing "freelance"; just put in the figures for the FULL SIZE SPRING, work out the size, and then reduce the diameter to scale.

The real problem comes when *making* the eye ends. Most practitioners machine up the eyes and then braze these to the top leaf. Fig. 46. This "works", but care must be taken. For the usual carbon spring steel the margin between the liquidus of the brazing alloy and the transformation temperature is only about 100 deg. C. (There is more margin, by the way, with hardened and tempered silver steel.) You must use the cadmium-bearing alloy AG2 (Easy-flo No. 2, MX12, or FSB2) and heat to no higher temperature than is necessary to get proper penetration of the alloy. (The liquidus is at 617/620 deg. C.) Apply the

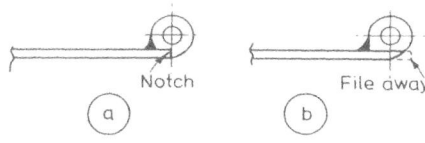

Fig. 46

flame to the eye rather than to the leaf, and have a trial run on a dummy if you haven't made up a master leaf this way before. It may help to use silver-brazing paint, but this is pretty expensive these days, and hardly worth buying specially. But it IS worth having a few lengths of fine-gauge brazing wire – I use No. 20 or 22 s.w.g. for this sort of job (0.75 to 1mm).

The alternative is to machine the top leaf from the solid and then heat treat. This is not as difficult as it sounds for leaves under about ⁵⁄₁₆in. wide provided you have a machine vice with good jaws – though I usually do the job by soldering the spring bar to a piece of mild steel first. I chew out the greater part of the material with a "Millencut" file first (quicker than milling!) and then mill down to about 2 mil, over thickness, the final shaving being taken off with a fine file to get a good finish. The ends are then shaped and drilled, the leaf curved, and then heat treated. About 5 minutes at just hotter than cherry-red (about 800 deg. C) – or 780 deg. C for silver steel – and then temper at *no less than* 310 deg. C (Blue). In fact, with silver steel and chisel steel a temper at "Gray" (350 deg. C) can be an advantage. The temper temperature also should be held for about 5 minutes, and for those without a salt bath the use of a heated sand-bed is advised. To reduce scaling, coat the spring with a paste of powdered chalk and water, and use a separate uncoated piece to judge the temperature.

The Heretical Method

This is for those who have Aga or Rayburn cookers, or a muffle furnace and, in effect, is going back to the Middle Ages! You carve out the spring as just described from *mild steel,* and then convert it to spring steel! We use the case-hardening or, rather, carburizing process. If you look back to Fig. 33 (page 48), you will remember that the "working stress" of about 70,000lbf/sq.in. applies only to the outermost fibers. When you get down inside the leaf to around a quarter of the thickness, the stress will be halved, and that is just below the yield point of mild steel. So, if we can convert just a little more than this thickness to high-carbon steel, and then heat treat it, the leaf will "behave". In fact, with these thicknesses it is easy to overdo it, and once the case has reached the center, the surface may reach too HIGH a carbon content!

The drill is simple. Make the spring, not forgetting to put the holes and the curve into it, from mild steel. (Don't drill the center hole until you have formed the curved shape, as this may cause a kink during the bending.) If you use BDMS, you must machine all over, to remove the stressed outer skin. Set a bolt through the holes in the spring eyes, to avoid "casing" the interior, and then put the leaf (or leaves) into a stout tin filled with Kasenit. Seal the lid with fire clay (fire cement will do, but will set very hard) to prevent air getting in. Set this in either a muffle heated up to 850-900 deg. C, or in the heart of the Aga firebox, and leave for 1½ hours – timing from when the box has reached fire heat. This should give about $1/64$in. (0.4mm) depth or a little more, but make a test first – comments below will explain why! The next step is "optional". If you can get the tin open *quickly,* do so, and quench the leaves in oil. This is not essential but does refine the grain of the uncarburized core a little. Now reheat the leaf to the usual hardening temperature 770/790 deg. C and quench in water or oil. Finally, temper at 330/350 deg. C – gray.

This process does work, and works well. The only problem may be in machining leaves to less than about $3/64$in. or 1.25mm thick. Even here, I have found that soldering the part to a block which can be held in the vice is satisfactory *provided that you remove the solder before carburizing.* But if using BDMS, you *must* machine – and fine finish – the back before soldering this down and machining between the eyes. Using the firebox of a stove does mean some careful attention to the fire, but that is all. (Heating at up to 930 deg. C does no harm, but may give too thick a case.)

Fig. 47 shows an experimental spring made this way, 2.75in. long (between centers of the eyes), 0.063in. thick and 0.25in. wide. It was machined from ¼in. square bar as described above, the eye holes being drilled in the square bar as first operation. After polishing and forming the curve (no center hole was drilled), it was set in an old typewriter-ribbon tin packed tight with Kasenit No. 1 and carburized for 2¼ hours at 890 deg. C (see on page 73). After cooling, the hard spring was cleaned and anti-scale paint applied, when it was heated to 790 deg. C for 10 minutes and quenched in water. It was finally tempered at 400 deg. C for 12 minutes – rather higher than suggested above. I had expected a "case" about 0.020in. thick, but the test piece set in the box showed that it had carburized right through, and was probably higher in carbon % than anticipated. Carburizing for 1½ hours would have sufficed – see on page 73.

The spring was then tested by applying a force to cause deflections corresponding the stresses of 70,000, 80,000, then 100,000, 120,000 etc. lbf/sq.in., the load being released between each application to check for permanent set. Fig. 48 shows the condition at 80,000lbf/sq.in. Failure occurred at something over 150,000lbf/sq.in. – Fig. 49. This would give a safety factor at the normal working stress of 70,000 of about 2.15.

I had hoped for an ultimate stress of around 160,000lbf/sq.in., and, as suggested above, it would seem that I had underestimated the activity of Kasenit No. 1 – which is really intended for "open hearth" carburizing. However, the experiment shows that the procedure is valid, and only needs a little experiment to establish properly. I would suggest that those having the necessary facilities try (a) carburizing for 1½ hours and/or (b) using a 50-50 mixture of fine charcoal and Kasenit as the packing medium, (c) quenching in oil. The procedure was rather tedious for a single spring, but a set of 6 or 8 can be gang-milled and all carburized at once. Previous springs which I have made this way have, of course, been of the "awkward shape" variety rather than for locomotive suspension!

I will conclude this section on leaf springs with a final comment – on "usage" rather than manufacture – to emphasize a point made earlier about lubrication. Those who are old enough to have driven cars in the "Vintage" era (pre-1930, for the record) or who have owned one subsequently, will appreciate the importance of this. Model springs should be assembled with a lubricant,

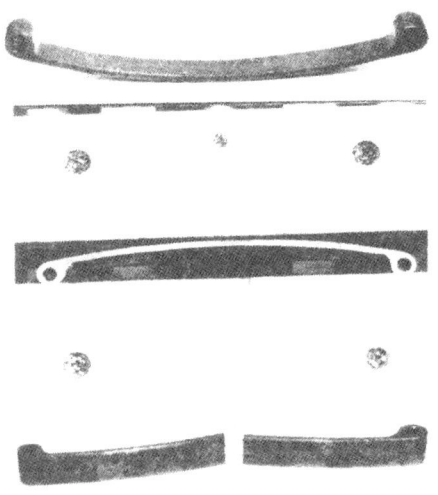

Top, Fig. 47 *The experimental casehardened spring after tempering.*
Center, Fig. 48 *The experimental spring loaded to a stress of 80,000lbf/sq.in.*
Bottom, Fig. 49 *The sample broke at a stress of just over 150,000lbf/sq.in.*

and be relubricated at intervals. I am a firm believer of *graphite* for this purpose, despite the claims for MoS_2, and use a graphitized "running-in compound" – still available from Halfords etc. – let down with thin oil; either Tellus 11 or something like "Three in One." For assembling heavier springs, I either coat them with the running-in compound neat, or use ordinary oil and sprinkle powdered graphite onto the leaves. When relubricating, try to unload the spring completely and open the leaves with a strip of thin steel or feeler strip, to encourage the oil to penetrate.

CHAPTER 9

Torsion Springs

Readers of my age will remember the days when clockwork toys had *coil* springs which were wound up, instead of the clock type, but they do have a raft of other uses – click springs, holding shut the lids of lubricators and so on. Then there are springs which are plain bars in torsion or even tubes (or both). All have occasional use on models and many applications in experimental gear. We will start with the first type.

The coil spring in torsion
These may range from a single coil (or even part of one) to long affairs, but the behavior is the same. Look at Fig. 50. AA is the axis of a coil spring, and we have applied a force P at one end, acting at right-angles to the axis. This will be resisted, if the spring is to stay put, by an equal and opposite reaction at R. The torque applied to the spring will be P × r, where r is half the diameter of the coils. (I have drawn these as a single line, as you see.) If you think for a moment you will see that if P = R, then there must be a uniform force equal to P acting on every coil, right down the length of the spring. Let us see what effect this has on a single coil – Fig. 51. AA is the coil axis as before. The force P and its reaction R will impose a *bending moment* at the point X, this bending moment being equal to P × r – numerically equal to the twisting moment or torque applied to the spring. This is the important distinction between these and the coil springs we

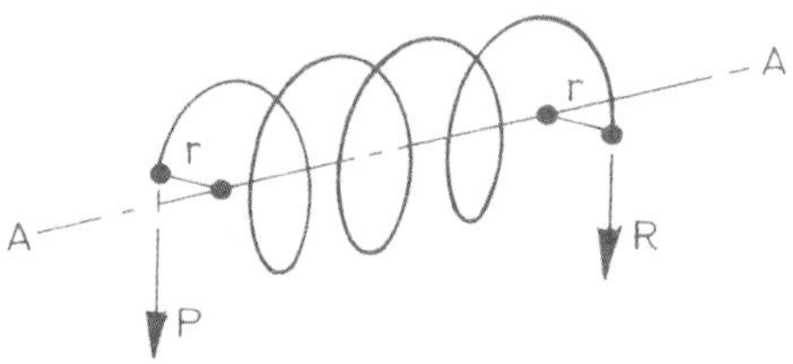

Fig. 50 *The helical torsion spring.*

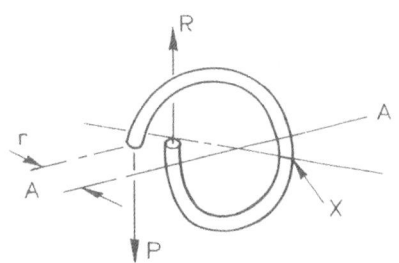

Fig. 51 *Loading conditions on a torsion spring.*

74 SPRING DESIGN AND MANUFACTURE FOR HOME MACHINISTS

have seen before – the stress is a BENDING STRESS. There is also a small direct TENSILE stress as well, due to the pull of the force P acting on the area of the wire, but this is usually small enough to be neglected.

Again, I won't go into all the math of the thing, but the bending stress is easily found from:

for round wire

$$f_b = \frac{32P \times r}{\pi d^3} \qquad (17)$$

and for rectangular wire

$$f_b = \frac{6P \times r}{b \times t^2} \qquad (18)$$

Where P = force, as Fig. 50
r = D/2 (D = coil dia.)
d = wire dia.
b = axial width of rectangular wire
t = radial thickness, ditto,
f_b = bending stress

If P is in lbf and other dimensions in inches, the stress, f, will be in lbf/sq.in. If P is in Newton and dimensions in millimeters, f will be in N/sq.mm. I need not remind you that for *square* wire b = t!

It is, perhaps easier if these two formulas are rearranged to give the *torque* or twisting moment, as under (T = P × D/2):

$$\text{For round wire .. } T = \frac{f_b \pi d^3}{32} \qquad (19)$$

$$\text{For rectangular. } T = \frac{f_b b t^2}{6} \qquad (20)$$

T being in lbf/in. or Newton/mm according to the units used.

Fig. 52: *A typical specially made torsion spring.*

The direct stress due to the "pull" on the wire is easily calculated as LOAD/AREA, which comes out as $4P/\pi d^2$ for round wire, and P/bt for rectangular. From what we saw when considering leaf springs, it is clear that this stress must be *added* to the bending stress on the outside of the wire, where f_b is tensile, but *subtracted* on the inside, where bending stresses are compressive. We need not bother about the latter, as the effect is to reduce the working stress. As I have already said, this factor is not of great importance so long as the D/d ratio is reasonable – more than 10 – and the working stresses given in Table V include an allowance anyway. But if the wire diameter is large and the coil diameter small, unavoidable in some cases, then the two stresses must be worked out and added; or, alternatively, work out the direct (tensile) stress first, and subtract this from the working stress f_b when using expressions (19) and (20). Working stresses for the usual materials are given in Table V.

Table V		
Mild Steel	35,000lbf/sq.in.	
H.D. Spring Steel	100,000 "	0.15in. dia.
	150,000 "	0.015in. dia.
Music Wire	130,000 "	0.125in. dia.
	150,000 "	0.062in. dia.
	170,000 "	0.031in. dia.
Stainless Steel	100,000 "	Very variable. Consult maker's data sheets.
Phosphor bronze	35,000 "	
Spring-brass	25,000 "	
Monel metal	40,000 "	Cold-rolled sheet.
Except for monel, these figures are for *wire*. See Table III, page 51 for leaf spring materials.		

These stresses are suitable for springs under steady load, or on which the load varies relatively seldom. The wind-up spring of a toy, for example, or that on a mousetrap! For springs suffering repeated loading, and which must "live" for thousands of operations, then fatigue life must be considered. This is a complex subject, and the easiest way for us to deal with it is to use a lower stress, so that the fatigue limit is unlikely to be exceeded. One authority recommends a reduction of the working stress as given above by 40%. The only disadvantage of this procedure is that the spring may be heavier.

Deflection
As each coil – indeed, each part of each coil – carries the same torque and suffers the same stress we would expect the deflection – the "twist" – in the spring to depend on the number of coils, and this is in fact the case. This deflection can be derived from the bending formula, but again I won't go into the details – the more so as it would give the answer in "radians" whereas we usually need degrees or number of turns. The expressions simplify to:

Round Wire:
$$\S = \frac{10 \cdot 18 TDN}{Ed^4} \text{ Revolutions} \quad (21)$$

or
$$\S = \frac{3665 TDN}{Ed^4} \text{ Degrees}$$

Rectangular Wire:
$$\S = \frac{6TDN}{Ebt^3} \text{ Revolutions} \quad (22)$$

or
$$\S = \frac{2160 TDN}{Ebt^3} \text{ Degrees}$$

N = No. of Coils
E = Young's Modulus
  Steel = 30,000,000
  P.B. and Brass = 15,000,000
  Other symbols as before

Although these two formulas are fairly easy to use given a pocket calculator, I show in Fig. 53 a Nomogram, due to the late Prof. J. B. Peddle, which may make matters a bit easier for those who find scale-rules handier than a mass of pushbuttons! Unfortunately, it was devised for fairly hefty springs, and though I have extended it a little it will not deal with wire smaller than about 21 s.w.g. The instructions are shown in the caption to the chart, but I will deal with its use a little later.

"Power" of a torsion spring

This is an unfortunate term, as it does not mean what it says! "Power" is a *rate* of doing work or releasing energy and includes an element of *time*. But you can let the same spring "run down" quickly or slowly, depending on how you use it; an 8-day clock spring usually develops a few milliwatts, but if you release it from the casing fully wound, it can develop a few lethal kilowatts! What is really meant when talking of "power" is "Energy", or "Resilience" – I prefer the former term. This is simply the product of the torque and the twist or angular deflection of the spring, following the well-known expression . . .

$$\text{Workdone, } W = 2\pi T.n,$$

where n is the number of revolutions made by the spindle. (Note that n is NOT the number of coils in the spring.)

The *units* of work – inch lbf or mm Newtons – will be the same as those for the torque, T. The actual *power* which the spring can develop, in HP or Watts, depends not on the spring itself, but on how fast you let it run down. This, in turn, depends on the mechanism to which the spring is applied, and that lies outside the scope of this discourse.

Example

A return spring is required to develop a torque of 3lbf/in. after 3 turns of the mandrel, which is ¾in. diameter. Find the *size of wire and the number of coils, if the working stress is 120,000lbf/sq.in.*

From Eq. (19), inverted to give d,
$d^3 = 32T/\pi f_b$
$= (32 \times 3)/(\pi \times 120,000)$
$= 0.000\ 255$
Take cube root to give $\underline{d = 0.0634}$ about 16's gauge = *0.064*.

To find the number of coils, invert Eq. (21)

$$N = \frac{\S \times Ed^4}{10.18TD}$$

$$= \frac{3 \times 30,000,000 \times (0.064)^4}{10.18 \times 3 \times 0.875}$$

(D = ⅞in. to allow for the wire diameter, and
$0.064^4 = 0.000\ 0168$)

N = *56.6, about 57*

Check this for yourself using the nomogram, Fig. 53 as follows.

Join values of "Modulus (E)" and "Coil Dia." and mark line "A". Join "3" on "Twisting moment" line with 0.064 on wire line (use the correct side) and mark line "B". Join marks on "A" and "B" and mark line "C". From "3" on angle of twist (Turns) line join through this mark, to read 59 on "Coils" line. This is as close as can be expected in view of the reproduction processes involved – the two figures agree to ± 1½%.

You can of course, work the other way round, knowing the desired torque and the available wire and space (which may govern the number of coils), so let us check back on the example given by Prof. Peddle when he published the Nomogram. (In 1913, by the way, so that the stress may be a little conservative.)

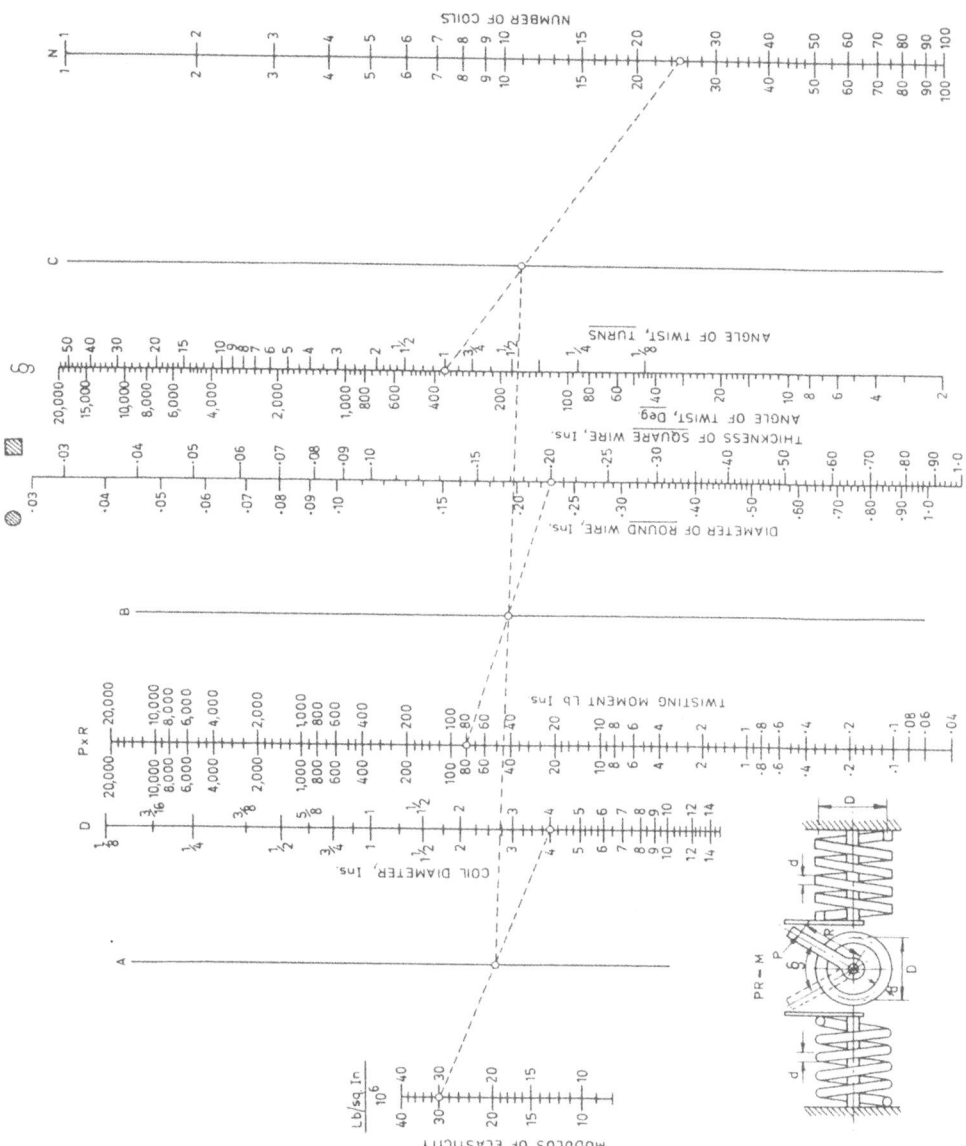

78 SPRING DESIGN AND MANUFACTURE FOR HOME MACHINISTS

Opposite: **Fig. 53 Deflection of helical springs in torsion.**
Connect the modulus of elasticity with the coil diameter and note intersection with axis A; connect the twisting moment with the thickness of wire and note intersection with axis B; join the intersections and extend the line to intersect axis C. A line through this intersection will connect the angle of twist with the required number of coils. The example shows that a square wire of steel, having a modulus of elasticity of 30,000,000, measuring 0.2in. on the side, wound into a spring of 4in. mean diameter and 25 coils, will twist one full turn under a twisting moment of 80lb./in. Note: This chart does not include the correction factor referred to below.

Use Eqn. (18), remembering that for square wire b = t

$$f = \frac{6 \times T}{t^3} \quad (T = P \times r = P \times D/2)$$

$$= \frac{6 \times 80}{0.2^3}$$

$$= 480/0.008$$

$$= \underline{60,000 lbf/sq.in.}$$

As suspected, this is on the low side – try working out for yourself the dimensions of a spring for the same duty, still using square wire, but working at a stress of 90,000lbf/sq.in. (Noting, of course, that there is no *harm* in a low stress, except that it makes the spring rather heavier.)

Correction Factor
The expressions used so far take no account of the fact that the wire is bent not as a beam but as a coil. This has little effect on springs where the D/d ratio is relatively large, as the inevitable tolerances on both wire and coil diameter – to say nothing of the behavior of the coils as the spring is wound up – introduce an effect which overwhelms any error due to curvature. However, when the D/d ratio gets below about 10/1 – as in the case of the spring in Fig. 54 – the curvature effect should be allowed for. This is, fortunately, very easy; the effect is to increase the stress found in Eqns. (17) and (18) by a factor "k", or to reduce the torque given by Eqns. (19) and (20) by the same factor. The deflection Eqns. (21) and (22) are unaffected, as these take account of the *actual* stress already. Values of k are given in Table VI.

As we normally use the equations to find the wire diameter to provide a given torque at a fixed working stress, and as d is proportionate to the cube root of f × k, the maximum effect on the spring design, even when D/d is as low as 3, is to increase the wire diameter by 10%. In fact, as easy a way as any when determining wire diameters is simply to use Eqns. (17) and (18) as they are, and then to multiply wire diameter d by $\sqrt[3]{k}$.

Table VI								
Ratio D/d	3	4	5	6	7	8	9	10
k, Rect. wire	1.29	1.20	1.15	1.12	1.11	1.09	1.08	1.07
k, Round wire	1.33	1.23	1.18	1.14	1.12	1.10	1.09	1.08

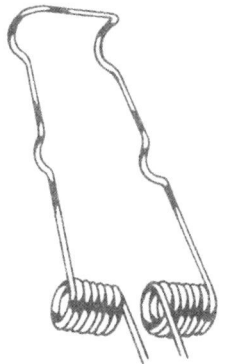

Fig. 54 *Correction factors must be applied when coils are as tight as these.*

This simplifies matters, as we do not really know the D/d ratio until d has been determined. This method should be applied to the Nomogram, Fig. 53, as it does not include the correction factor.

Practical aspects of torsion springs

The actual winding has been covered already – though a spring such as that in Fig. 54 may tax your ingenuity! Square or rectangular wire is more economical and, for a given duty, occupies less space but the material is not easy to come by, unfortunately. The mandrel on which the spring works is important; any but short springs (less than 6 coils) may, and any with more than 10 coils will, need some support, otherwise the spring may cockle as it is wound up and give an irregular torque as it unwinds. It is best to guide using an internal mandrel, and this *must* be smaller than the inner diameter of the coil, as the coil diameter diminishes as it is loaded. It is essential, too, that the spring does not bind on the guide mandrel when fully wound up. Fortunately, the clearance required is easily determined. Suppose we have a 40-coil spring which is to be wound up by ten turns. There will then be 50 coils. The maximum guide-mandrel diameter must be not greater than 40/50 or 0.8 × the inner diameter of the unloaded coil. If, however, the guide is in the form of a tube outside the coils, then the ID of the tube would have to be just a little larger than the OUTER diameter of the unloaded spring. In practice, I would advise an extra clearance of between ⅕ and ¼ of the wire diameter as well.

It is fairly important that springs such as these should be wound UP, not DOWN – that is, such that the number of coils increases – and that they should, where possible, have a small tension in them when at minimum load. The internal stressing is complex, and if loaded by *UN*winding the locked-up stresses incurred when first making the spring can cause the total stress to exceed the safe maximum. It is also important to ensure that the spring cannot bind endways when fully loaded. In the case already cited above, there must be adequate end-room for *50*coils, not the 40 originally wound, and with a small clearance between coils even then. A coil-to-coil clearance, fully wound, of 5% of the wire diameter should be aimed at.

End fixing needs careful thought. Where possible the wire should be arranged as at (a) in Fig. 55, rather than (b) – the latter can impose a fairly high bending stress in the coil where the "ear" is formed. The tangential part of the fixing in (a) can be short. However, (b) is much better than style (c), where the fixing lug is bent inwards, as (c) imposes an even greater stress at the junction of lug and coil. If the wire end can be clamped instead of forming a hook, as at (d), this is almost ideal, as there is no

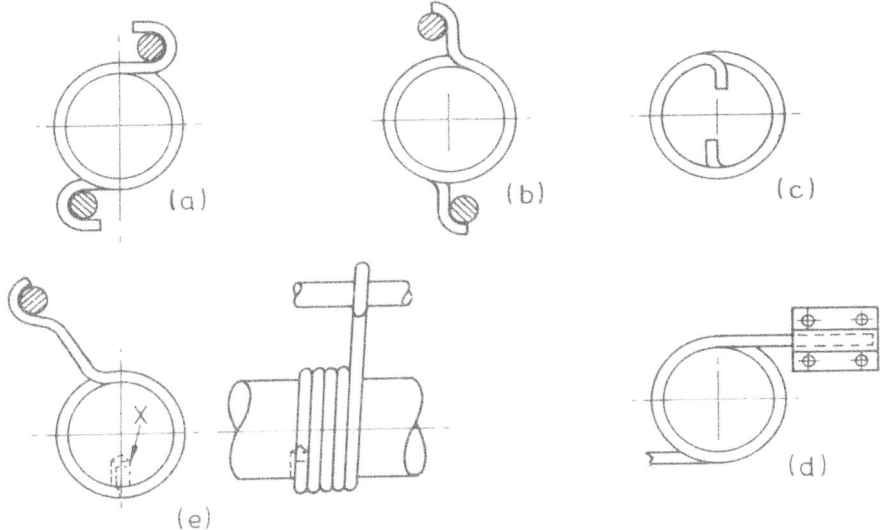

Fig. 55 *Termination of torsion springs, see text for comment.*

increase in stress at all. The wire can, of course, be hooked over the end of the clamp for safety should the clamp work loose. The arrangement at (e), where the end of the spring at "X" is bent over and caught in a drilled hole in the shaft, needs special care, as the small lug is now in *shear,* and unless the edge of the hole is radiused, there will be an acute stress-raiser at this point. It is a common arrangement, but does require care both in design and construction.

The Flat (Spiral) Torsion Spring
This is shown in Fig. 56. The coil may be of round wire or rectangular (flat) strip, and may be fixed at 0 and deflecting from A or vice versa. This configuration is rather complex, as instead of finding a uniform bending moment throughout there will normally be a maximum at the point X and a minimum at the center. However, there *can* be a high bending moment at the center 0 if the coil is very tight at this point. So far as strength is concerned, we need bother only with the maximum, M = P × r, and Eqns. (17) and (18) apply (see page 75). These can be rearranged in terms of torque, T = P × r, exactly as before, using Eqns. (19) and (20). So far, so good.

However, when we come to deal with

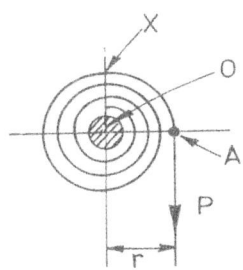

Fig. 56

the deflection – the number of turns (or degrees of twist) to provide this torque, things are rather different. Unless we use a fairly rigorous analysis, to allow both for change in bending stress *and* the variation in curvature as the center is approached, we must use some approximation. There is a number of these, but the one I use is:-

For Round Wire

$$\S = \frac{3.25\,PrL}{E \times d^4} \text{ Turns} \quad (23)$$

For Rectangular Section

$$\S = \frac{1.91\,PrL}{E \times bt^3} \text{ Turns} \quad (24)$$

L = total length of wire.
E = Young's Modulus [See (22)]
P and r as Fig. 56.

As before, consistent metric or imperial units should be used, with E in lbf/sq.in. or N/mm².

These equations will give reasonable results so long as the spiral is of fairly large mean radius, but for "tight" springs, the stress will be increased; this means that the safe value of P must be reduced – for example, by 10%.

The case of a *spring clock,* where the coils are close wound and the number of coils is large, is more complex. In addition, the material used for such springs can be worked at a very much higher stress – something like 200,000 lbf/sq.in. Finally, the user is concerned not with the maximum torque which can be developed, but with the total energy released when run down between this maximum and the minimum value needed to drive the mechanism. He must also allow for the losses due to friction between adjacent coils. It is very rare that even an experienced horologist will actually wish to *make* such a spring, and those who do so will have access to the specialist books on the subject. Readers who would like more information on the subject are advised to write to the British Horological Institute asking for a suitable reference – B.H.I., Upton Hall, Upton, Newark, Notts.

Torsion-bar Springs

While the torsion springs so far considered provide a relatively large rotation and relatively small torque, there are occasions where the reverse is needed. In such a case, the torsion bar provides the answer – Fig. 57. The bar or rod is firmly fixed to an anchorage at one end, A, and is carried in a bearing of some sort at the other, B. Any force applied to the lever arm will induce a twisting – shear – stress in the bar. The condition is exactly the same as that of a shaft carrying power, except that the angle of twist is much greater, and the same stress conditions apply. Thus: For solid shafts

$$f_s = \frac{16\,T}{\pi D^3}$$

or $\quad T = \dfrac{\pi D^3 f_s}{16} \quad (25)$

and $\quad \S_r = \dfrac{32}{\pi}\,\dfrac{TL}{GD^4}$ (radians)

or $\quad \S = \dfrac{584\,T \times L}{GD^4}$ (degrees) $\quad (26)$

$\S$ = angle of twist at free end.
f_s = safe shear stress. Lbf/in.² or N/mm².
T = torque, W × R, Lbf/in. or N/mm.
D = O.D. of shaft, in. or mm.
L = length of shaft, in. or mm.
G = Torsional modulus of rigidity.

For *hollow* shafts, of internal diameter d, (25) and (26) become:

$$T = \frac{\pi(D^4 - d^4)f_s}{16D} \quad (25a)$$

$$\S = \frac{584\,T.L}{G(D^4 - d^4)} \quad (26a)$$

Values of G are given in Table I, but the table is repeated below for convenience.

Table I (Repeated)		
Material	G.lbf/sq.in. × 1,000,000	G. Newton/sq.mm × 1,000
Carbon Steel	11.4	80
Piano Wire	12.0	83
18/8 Stainless	10.0	69
Phos. Bronze 70/30 H.D.	6.0	41
Brass	5.0	35
Monel	9.5	65

The safe shear stress will be the same as those given in Figs. 5 to 8, but the diameters of such torsion bars are usually rather larger than shown there.

Further, you may have to use material which has been neither hard drawn nor heat treated – though you can, of course heat treat silver steel yourself if need be. Suitable working shear stresses for such material are as follows:

Mild Steel	28,000lbf/sq.in.
0.4% Carbon Steel	45,000
ditto	55,000 Heat treated
Silver Steel	65,000 Tempered to 315°C
Bronze	22,000

Example
Calculate the necessary diameter and length for a mild steel torsion bar to carry a torque of 150lbf/in. at a deflection of 10 degrees.

For MS, f = 28,000 and G = 11.4 million.

From Eqn. 25, inverted,
$D^3 = 16T/(\pi \times f)$
$= 16 \times 150/(3.14 \times 28,000)$
$= 0.0273$

Take cube root for D
D = <u>0.301 inches</u>

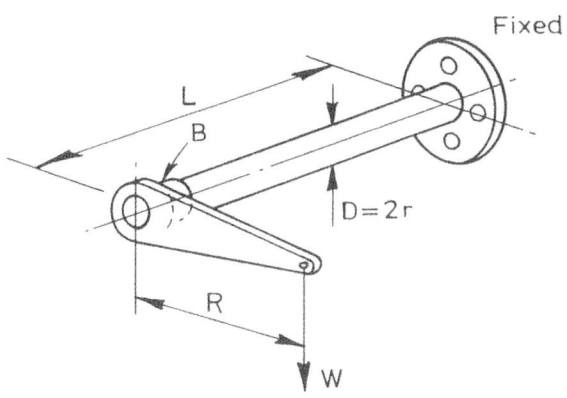

Fig. 57

From Eqn. 26 inverted

$$L = \frac{\S.GD^4}{584T}$$

$$= \frac{10 \times 11{,}400{,}000 \times 0.301^4}{584 \times 150}$$

$$(0.301^4 = 0.00821)$$

$L = \underline{10.68 \text{ inches}}$

Now work out for yourself the stress and the length needed if standard $^5/_{16}$in. (0.313) diameter stock has been used.

Next, let us try an example using *hollow* tube instead of a bar, taking the bore giving a small clearance on the above solid bar – for a reason which will appear shortly! Use exactly the same stress and loading, but find the O.D. of the tube if it is to be $^5/_{16}$in. (0.313) bore. This time, use Eqn. 26a inverted to find D:

$$D^4 - d^4 = \frac{584\,TL}{G \times \S}$$

$$= \frac{584 \times 150 \times 10.7}{11{,}400{,}000 \times 10}$$

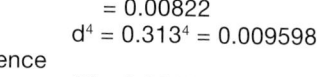

= 0.00822

$d^4 = 0.313^4 = 0.009598$

Hence

$D^4 = 0.00822 + 0.009598$
= 0.01782

and $\underline{D = 0.365(4)}$

(We should not be working to "tenths"!)

This gives a tube thickness of very nearly 22 s.w.g. Check for yourself the deflection if we had used $^3/_8$in. O.D. × 22 s.w.g.

The first point to notice here is that the hollow spring is much lighter; the solid bar is 2½ times heavier. More important, a thin tube like this withstands *fatigue* conditions much better than a solid bar. (I won't go into the reasons for this, except to say that the *stress* range – the difference between the maximum and minimum stresses in the metal – is much lower.) But that is not all. We can use the hollow spring in conjunction with a solid one to make the assembly much more compact.

The Concentric Torsion Bar
See Fig. 58. Here we have a solid spring inside a hollow one, securely united

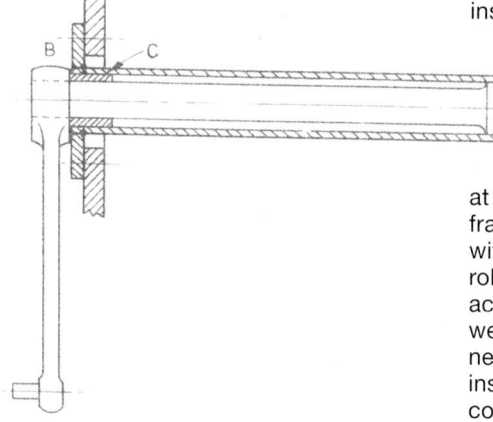

Fig. 58

at "A", the hollow part attached to the framework of the machine at "B" and with a simple bush (or it could be needle rollers for a "luxury" job!) at "C". The active length is L, and if the dimensions were those we have just calculated this need be only half that we found, 5.34in. instead of 10.68; the spring is more compact. Further, the whole spring can be made up as an assembly and inserted through the hole in the framework at "B", whereas the single torsion bar needs an attachment at one end and a bearing at the other. The first car I owned which had "Independent Front Suspension" (a 1939 Vauxhall "10") had springs of this type, though the arrangement of main and rebound springs, shock absorbers, etc. was rather more complex than my sketch!

In practice, the attachment at "A" must be secure; a cross-pin, key, or even splines, is not really good enough, and the joint should be welded or brazed. This may mean using unheat-treated steel, though with some have tempering temperatures high enough (EN32 is an example), the use of AG2 ("Easyflo" No. 2) silver-brazing alloy will not draw the temper. The inner bar would better be slightly enlarged at C, to accept the bush and make assembly easier. Finally, in designing the spring we should, after making a preliminary calculation as shown above, recalculate the tube dimensions and perhaps that of the solid bar as well, so that a standard diameter and gauge thickness could be used for the tube. Finally, there is no reason at all why the same stress, or even the same materials, should be used for tube and bar; all you have to do in such circumstances is to work out the diameters separately, for the same value of the length, L.

Odds and Ends

Before going on to the next section, there are a few "special cases" which apply to springs in bending, clips, and the occasional unusual conformation which may be found. Fig. 59 shows two forms of click or ratchet spring, where the bar has been tapered to increase the deflection. It is a variation of the "constant stress" spring, and may be straight as at (a) or curved as (b). The thickness, t_2, is determined to carry the shear load, such that $b \times t_2 = W/f_s$. For spring steel, or tempered silver steel, f_s may be taken as 65,000 to 75,000 lbf/sq.in. but lower stresses should be used if it is not heat treated. The thickness at the fixed end is found exactly as if it were a laminated spring with but one leaf – see the example applied to Eqns. (7) and (8), page 49. The deflection at the load point, W, however, cannot be found from the conventional method, as the "beam" is a tapered one. The analysis *can* be complicated, but reasonable results will be obtained if we write:

$$\text{deflection } \S = \frac{1.3 \, f_b . L^2}{E \times 1}$$

Where f_b = max. bending stress
L = actual length of spring
(= L_2 at Fig. 59b)
E = Young's Modulus
I = $b \times t^3/12$

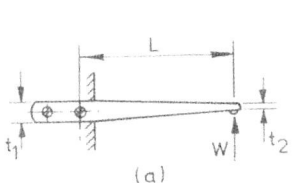

Fig. 59

Fig. 60

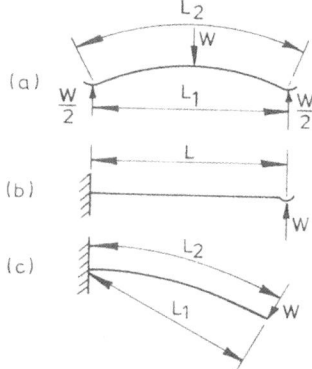
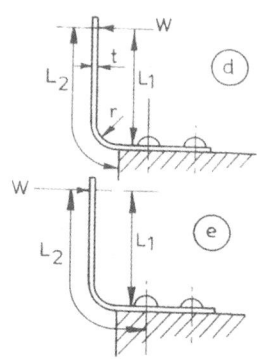

Note that the distance L_1 must be used to calculate the *stress,* but L_2, the actual length of the spring, is used when determining the deflection. The alternative method of treating case (b) is to calculate both "torque" and deflection as if it were a helical torsion spring with only *part* of a coil in action, but such click springs are seldom true arcs of a circle. (Their main application is to obtain sufficient flexibility in limited space.)

Fig. 60 shows a few "flat" spring applications. In each of the cases shown, the *stress* is determined (using the expressions already given for flat or leaf springs) based on the length L_1, but the *deflections* must take account of the actual metal length L_2 where this is different. However, where a sharp radius appears, as in cases (d) and (e) the stress here will be higher. The radius "r" should not be less than 2 × t (the thickness), but even then a reduction in working stress of about 15% should be allowed. At r = t the reduction must be 20% and for really sharp bends, 35%. Such sharp bends, however, are very likely to fail in the actual bending, and can only be achieved by bending before heat treatment.

Finally, note the effect of the *direction* of the load in sketches (d) and (e) shown by the arrow at W. The latter arrangement would be better if there were a clamping-piece between the screws and the spring.

CHAPTER 10

I.C. Engine Valve Springs

This section is not a dissertation on I.C. engines, for there are many other mechanisms in which a spring is used to control the behavior of linkwork or components operated by cams, but the valve spring is, perhaps the most common and in some ways the most difficult. There is no problem, of course, in designing the actual spring; the difficulty which most readers seem to experience is in deciding how strong the spring must be. The method described now is simple, and can be applied to any case where spring return of a mechanism is needed and to any shape of cam; in short, to any situation where an *acceleration* sets up otherwise unrestrained forces in the system.

Almost all shapes of cam can be analyzed mathematically, and for those with a taste for this sort of activity I give the procedure at the end of the section. But as we need a graphical display of the result anyway, and because graphical method enables us to see the effect of alterations to the cam-shape very quickly, this is the method I recommend – and, in fact, used when designing valve-gear was part of my daily work. In any case, if this cam shape is not a simple mathematical form – and those I was accustomed to seldom were – the equations can become very complicated.

The first step is to draw out the shape of the cam to as large a scale as possible

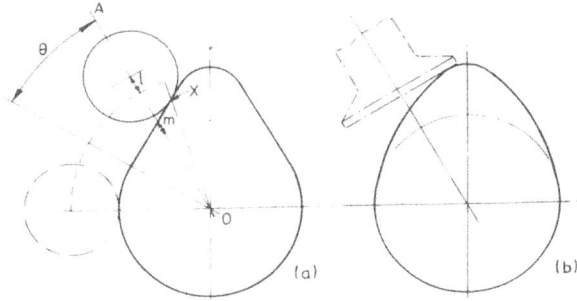

Fig. 61 *The effect of the cam follower. Although the roller lies θ from the start of lift, at **OA**, the actual contact lies further round at **OX**. The effect is even more pronounced when a flat follower is used as at (b).*

– 10 to 1 is convenient, but even larger should be possible for cams of model size. From this, the valve lift curve can be determined. There are two very important points to notice here. First, the shape of the cam *follower* must be allowed for – see Fig. 61. At (a), we have the usual roller-type, and you can see that although the cam has rotated only through the angle θ the roller is actually in contact with the cam further on, at X. The "lift" at this angle is the distance "l", between the center of the roller and the dotted roller base circle, NOT the distance "m". The effect can be even more marked if a flat follower is used, as at (b). The second very important point is that this lift must be translated to *valve* lift, by taking into account the lever ratio of any rockers, etc., between the cam and the valve itself. This is because the spring is at the valve, not at the follower end of the linkwork. (See Fig. 66 later.)

Fig. 62 shows the procedure. The lift has been measured at every 5 deg. (though only two such are shown), measuring between the roller base circle and the locus of the center of the roller. It is then multiplied by the lever ratio of the rocker-gear and plotted as seen in the upper diagram (A) of Fig. 63. Note, however, that the "degrees" have been converted to "time" – not forgetting, in the case of a four-stroke engine, that the cam rotates at *half* engine speed! if the engine makes 1,200 rpm, the cam makes 600, and 360 deg. occupies 1/600 minute or 1/10 second, so that 5 deg. takes 0.00139 seconds. This is a bit awkward, and what I do is to set out the time scale in milliseconds (0.001 sec.) and plot the valve lift on this. The timescale should, of course, be set up at the *maximum* speed at which the cam will rotate if this is variable.

Having plotted this curve, as at Fig. 63A we now convert displacement to velocity. Work out the velocity in *feet*/sec. or *meters*/sec. at this stage. Lift starts at "a" and at "b" the lift is "s". The *mean* velocity is thus s/ab – if ab is 0.001 sec., this will be 1,000s. From "b" to "c" the *additional* lift is "t", so the mean velocity over this period is t/bc. And so on, along the whole curve. I usually make a table of the figures, and then plot them on curve (B). Don't forget that it is the *increase* in lift between the points a-b-c, etc. which is used, *not* the total lift at each point.

Important note. The mean velocity between a-b must be plotted *midway* between a and b on the velocity curve, as shown at "d" and "g", etc., on 63B.

When this velocity curve is complete, do the same thing again; measure the *change* in velocity at each time-ordinate

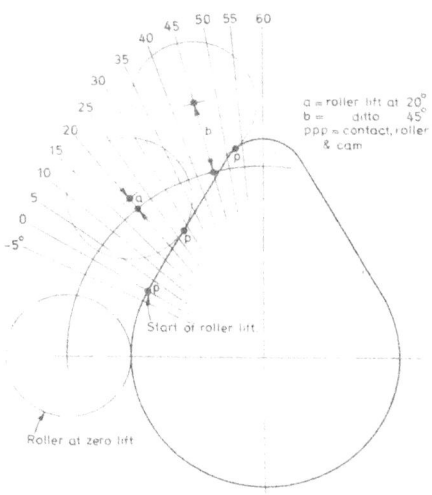

Fig. 62 *The method of plotting the lift curve from a cam. See text for procedure.*

(a, b and c again, but on curve "B") divide by the length of time – if 1/1,000 sec. then simply multiply by 1,000 – and plot the values of *acceleration* so found on curve (C) – again at the mid-points. These values will be in feet/sec.^{2} or meters/sec.^{2} and even on a small cam may be surprisingly large – possibly 1,000ft./sec.^{2} or more.

This does not take long, and you will notice that I have only shown half of the total cam period; in almost all cases, cams are symmetrical about the nose. However, if you have an unsymmetrical cam you must treat the whole lift period in this way.

You now have a graph of the acceleration over the whole of the lift period of the cam, and could, using this, calculate the forces similarly. Force = mass × acceleration. In Imperial units, mass is "weight/g" (W/32.2) and force will be in lbf. In metric units, the mass is in kilograms and the force in Newtons. Work in whichever is most natural to you! However, the actual design of the spring will be much simplified if you replot curve C, not on the basis of *time,* but on *valve lift* (not forgetting to correct for any lever ratio between cam and valve-head).

I have done this on Fig. 64. The lift at each ordinate – for example, each 1/1,000 sec. – is read off from Fig. 63a and the corresponding acceleration from Fig. 63c and then replotted. You will see that there is a noticeable difference in shape. You will also notice that there is a scale of *force* as well as of acceleration; this is made simply by multiplying the acceleration figures by the mass (W/g) of the total valve gear as seen at the valve. The curve is now one of FORCE vs. VALVE LIFT, the region marked (+) indicating force tending to hold the follower against the cam,

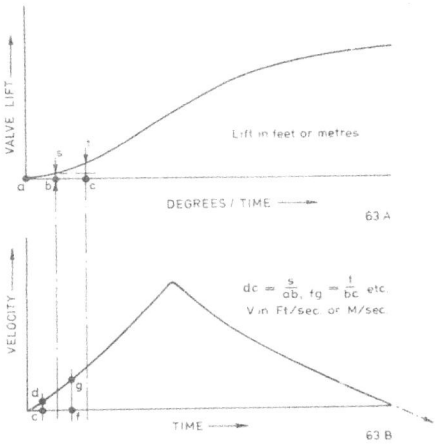

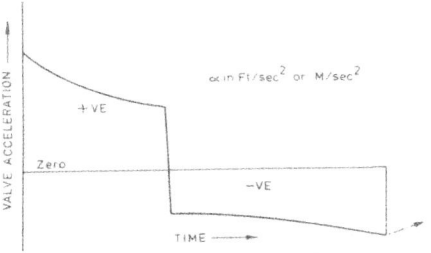

Fig. 63 *Graphical determination of valve acceleration from the lift curve.*
Note: These curves are illustrative only and are not to scale.

and that marked (–) needing a spring to counteract the forces. We can use this directly to determine not only the maximum spring force needed, but also the rate of the spring, and then use the nomograms I have already shown to design the spring.

The dotted line on Fig. 64 shows the first approximation to the spring force needed. I usually allow about 30% more

I.C. Engine Valve Springs 89

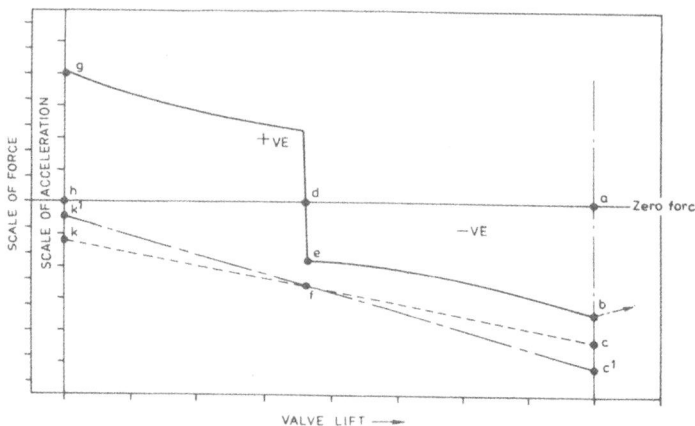

Fig. 64
Determination of spring rate. See text for procedure.

than the acceleration force at the salient points of the curve. This means that the engine can "run away" to about 15% overspeed before valve-bounce occurs – and, of course, if valve-bounce DOES take place this may help to prevent the engine from running even faster; a useful safety margin! So, bc = 0.3 × ab, ef = 0.3 × de, to give the straight line c-f-k as the spring force line. a-h is the valve lift, so the spring RATE will be (ac-hk)/ah lbf/in. or N/mm depending on the units. Note that [ac-hk] and ah must be figured in force and distance from the scales of the graph, of course.

However, there is another factor which I ought to mention, even though this is NOT supposed to be a course in engine-design! At "h", the valve is closed, and in the case of any throttle-controlled engine (gas or petrol), the spring force here must be enough to hold the exhaust valve closed against throttle suction. So, the *minimum* value of k-h must be 15 × the area of the valve port in imperial units, and 0.105 Newton × valve area in sq.mm in S.I. units, to be on the safe side. (To hold a poppet valve gas-tight, we usually reckon on about 9 to 10lbf/sq.in. on the valve-head. This is the minimum spring force needed on any poppet valve.)

The second point to notice is that the maximum load on the valve gear – pushrods, etc. – is the force g-k, the sum of the spring load and the initial acceleration force. This maximum *can* be reduced by stiffening up the valve-spring but reducing the initial force, as shown by the line c^1-f k^1 provided the condition just mentioned is met. Indeed, one can play about a lot with this curve by actually altering the acceleration curve to make it fit better an "easy" spring regime, for so long as the *areas* of the positive and negative regions remain the same on the time-based acceleration curve and the position of the zero crossover remain the same, the same maximum valve opening will result – though the *rate* of valve movement will change, of course. We then reverse the previous procedure, working from the acceleration to the velocity curve, and from velocity to displacement, to find the new cam shape.

Fig. 65 shows an actual cam and calculation sheet of this type, for a 2-

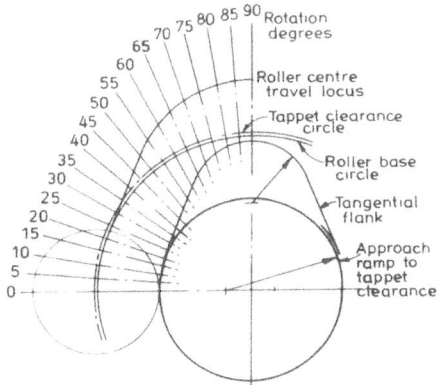

Fig. 65 An example of an actual spring force calculation, for a valve-in-head uniflow 2 stroke diesel engine. The valve is 1½in. dia and the lift 0.43in. Above: the cam, right: acceleration diags. (Reproduced, with permission, from Diesel Engine Design, by T. D. Walshaw, published by Geo. Newnes Ltd., 1950.)

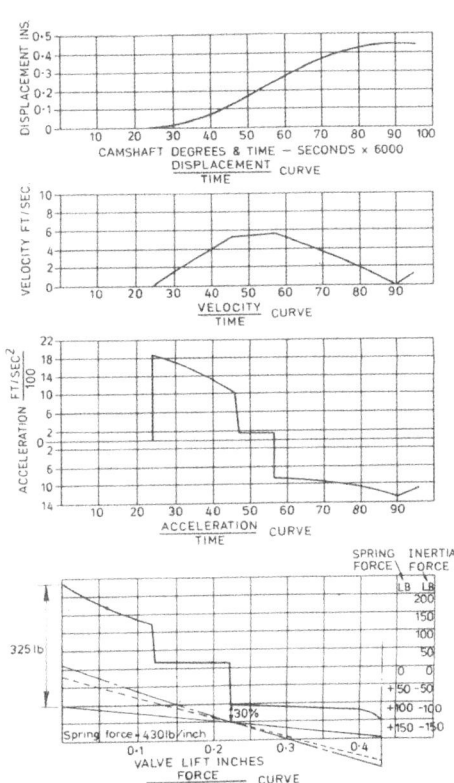

stroke uniflow diesel engine. On the lowest curve (d), the full line shows the 30% rule applied, giving a total gear loading of 325lbf. The chain-dotted line reduced this to 225lbf, but gave zero force with valve closed. The dotted line was better and provided the necessary "valve shut" force. However, as the design developed the high initial acceleration (1800ft./sec.) was reduced and part of the constant acceleration period eliminated, simply to give lighter loads on the valve gear. (With this particular type of 2-stroke, a very fast rise-time was needed to achieve the desired scavenge ratio but, as always, compromise between the ideal and the practicable had to be made!)

Valve Gear Mass
To find the forces, you need to know the weight (mass, to be accurate) of the rockers, pushrods, cam-followers, etc. In the case of model engines, it is difficult to *calculate* these accurately, so they must be weighed. However, there is a slight problem with rockers, as these are *rotating* parts. Those who know about such things as radii of gyration and polar moments of inertia can cope; the others will find that to take one-third of the total weight of a double-ended rocker as acting at the valve stem is a reasonable approximation.

In some books, you may find that the effect of the valve rocker of unequal ratio is quoted as being the *square* of the ratio y/x in Fig. 66. However, if you follow the procedure which I have outlined, and do all acceleration calculation in relation to *valve* lift, acceleration, etc. It is only necessary to multiply the mass of the pushrod, etc., by the simple ratio y/x. The effect of the lever ratio on the acceleration of the parts has already – and automatically – been allowed for.

Spring Vibration
This can be a problem, and not only on I.C. engine valve-springs, but I have heard of no serious difficulties with the very small springs used in models. However, for what it is worth, the approximate natural frequency of a coil spring (compression or tension) is given by:

$$F = \frac{12{,}700\,d}{D^2 \times n} \text{ cycles/sec. (Hz)}$$

D = coil dia., in.
d = wire dia., in.
n = no. of coils

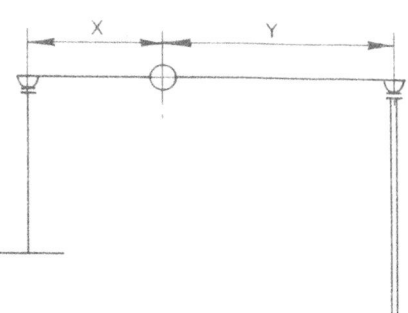

Fig. 66 *The lever ratio of any rockers must be taken into account when calculating accelerations and spring forces.*

This is necessarily *only* an approximation, but in any case, surging is usually found only at fairly high harmonics of this frequency and often a change in spring dimension only brings another harmonic into play. (Increasing the rate increases the value of F.) If serious trouble is experienced the best solution is to use *two* valve springs, wound to opposite hand, one inside the other. Or to fit a separate spring at the cam follower.

Analytical Method
Fig. 67 shows the mathematical expressions for acceleration for the two most common forms of engine cam. These look rather formidable but can be "programmed" into the more sophisticated pocket calculators or, of course, into a personal computer. The results should be tabulated at intervals of 5 deg. or so, converted to valve lift and then plotted to obtain a diagram as Fig. 64. After that, the procedure is exactly the same. The main objection to this method is that while it may be marginally more accurate, it is very difficult to work back to a new cam profile should it be found desirable to alter the acceleration curve.

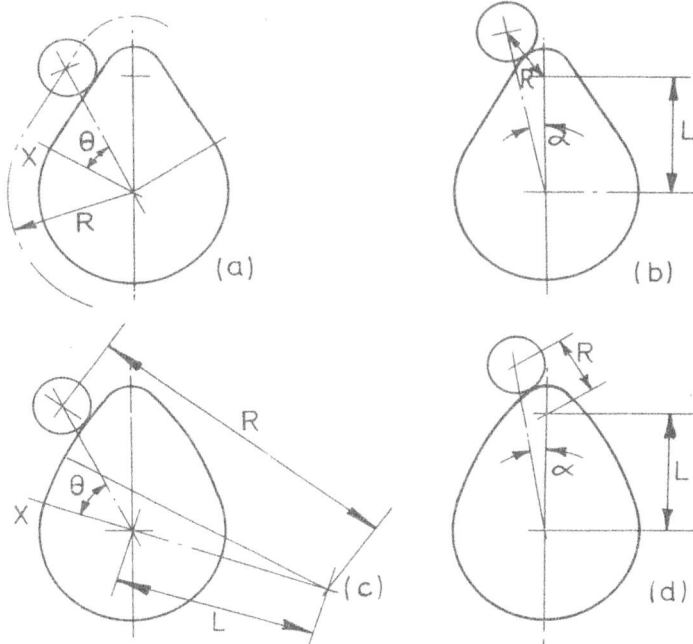

(i) *Tangent Flanks, Diagram* (a)

$$f = \omega^2 R \left(\frac{1 + 2\tan^2\theta}{\cos\theta} \right)$$

(ii) *Curved Flanks, Diagram* (c)

$$f = \omega^2 L \left[\cos\theta + \frac{S^2 \cos 2\theta + \sin^4\theta}{(S^2 - \sin^2\theta)^{1.5}} \right]$$

(iii) *Cam Nose, Diagrams* (b) and (d)

$$f = \omega^2 L \left[\cos\alpha + \frac{S^2 \cos 2\alpha + \sin^4\alpha}{(S^2 - \sin^2\alpha)^{1.5}} \right]$$

Where f = acceleration, ft./sec.^{2}
R = radius, ft.
ω = angular velocity, rad./sec.
   $= \dfrac{\text{r.p.m.} \times 2\pi}{60}$
L = length, ft.
$S = \dfrac{R}{L}$
$\left.\begin{array}{l}\theta\\\alpha\end{array}\right\}$ = angles in degrees
X = start of lift curve

Fig. 67 *Acceleration equations for two types of cam. Note that equation (iii) relating to the cam nose is the same for both types. (Reproduced, with permission, from Diesel Engine Design, by T. D. Walshaw, published by Geo. Newnes Ltd., 1950.)*

CONCLUSION

We have, I think, looked at most of the conformations of spring likely to be needed, but inevitably some special cases have been left out – circlips, belleville washers, and others. However, I hope that what *has* been included will enable you to get to grips with the ordinary problems and, perhaps, to make an approximation to the solution of those which are out of the ordinary. However, I must repeat the sense of what I said at the beginning. Tolerances on wire diameter, and uncertainties in manufacture – to say nothing of variations in material – do have an effect, and the smaller the spring the greater the variations from calculated results are likely to be. For all that, I think you will find that time spent in "working out" before you start will be a considerable saving in time overall, for at the very least it gives you some idea of what size to make your initial experimental spring!

Index

Acceleration, Valve Gear	89, 91, 92
Accuracy of calculations	33, 66
"ACRU" spring winder	39
Bending Stress	
leaf springs	48, 51
torsion coil springs	76
Cam lift/acceleration	88, 93
Casehardened springs	73
Charts	
Clamping leaf springs	61
Coil spring design procedure	29
rate	25, 26
torsion	74
winding	35
working loads	20–23
Correction factors	79, 88
Correction factors, coil springs	9
Coil spring, loads	20–23
deflection/rate	25, 26
Leaf springs deflection	60
design factor "K"	57
leaf thickness	59
Rate, tension/compression springs	25, 26
Spring back	37
Torsion coil spring design	78
Working loads, coil springs	20–23
leaf spring	59
Wire, safe stress	12–15
Deflection, coil springs	7, 25, 26
laminated springs	52, 53, 60
torsion springs	76, 78
Design procedure, coil springs	29
Diameter ratios, effect of	16
Examples, worked	30, 64
Hook ends, tension springs	43
Initial tension	42
Laminated spring – see "Leaf" Leaf length	70
Leaf springs	
bending stress	48
clamping	61
compound	62
deflection	49, 53, 60
design	55

formulas summary	54
manufacture	69
principles	47
tapered	52
Load, coil springs	20, 23
Locked-up stresses	36
Lubrication	73
Materials, strength	10, 48, 51, 76
Mild steel, use of	67, 72
Modulus of elasticity	12, 51
Music wire gauges	19
Nomograms. See Charts	
Piano (Music) wire	10, 19
Power of a spring	77
Rate	
defined	8
adjustment	44
coil springs	25, 26
leaf springs	55
Scale springs, difficulties	8, 61
Shackles	70
Slotted leaves	66
Spiral spring	81
Spring back	35
Stress checking	22
Stresses, working	
bending	48, 76
shear (in wire)	12–15
S.W.G. numbers	19
Tangent cams	93
Tension, initial	42
Terminations, coil springs	43
Tolerances on design dimensions	33
Torsion bar	82
Torsion coil spring	74
Tufnol, use of	62
working stress in	51
Valve-gear, effect of	91
Valve springs	87
Vibration in springs	92
Winding coil springs	35–46
Wire	
gauges	19
materials	10
rectangular	45
safe stresses	12–15
Worked examples	30, 64
Working stresses	12, 48, 76

Index 95